P9-CKF-394

# FLY-FISHING THE 41ST

Also by James Prosek

*Trout: An Illustrated History*

*Joe and Me: An Education in Fishing and Friendship*

*The Complete Angler:*
*A Connecticut Yankee Follows in the Footsteps of Walton*

*Early Love and Brook Trout*

# FLY-FISHING THE 41ST

*Around the World*
*on the*
*41st Parallel*

JAMES PROSEK

HarperCollins*Publishers*

HarperCollins books may be purchased for educational, business, or sales promotional use. For information, please write: Special Markets Department, HarperCollins Publishers Inc., 10 East 53rd Street, New York, NY 10022.

FIRST EDITION

Designed by Elliott Beard

Printed on acid-free paper

Library of Congress Cataloging-in-Publication Data
Prosek, James, 1975.
Fly-fishing the 41st: around the world on the 41st parallel / James Prosek.
p. cm.
Includes bibliographical references.
ISBN 0-06-019379-4 (hardcover)
1. Fly fishing. 2. Prosek, James, 1975—Journeys.
I. Title: Fly-fishing the forty-first. II. Title.

SH456.P765 2003
799.1'24—dc21 2002069059

03 04 05 06 07 /RRD 10 9 8 7 6 5 4 3 2 1

# Author's Note

The idea to travel a latitude line around the world was my editor's, Larry Ashmead at HarperCollins. To travel my home latitude was my agent's, Elaine Markson. The idea to live and raise his son in Easton, Connecticut, on the 41st parallel, was my father's. To infect me with a love of fish—I'm not sure who's responsible for that. These factors determined the route of my trip.

# PART I

# The 41st Parallel

One day, I left in a straight line from home at 41 Kachele Street, east along the 41st parallel, following my passion for fish. It was a journey not only away from home, but toward it; which is the beauty of traveling in a circle, and the irony of adventure. This suited me, for in the event that I strayed—as I would likely take some latitude with the latitude—as long as I could find my way back to the 41st parallel I would not get lost.

My home latitude, 41°N, contained along its length some of the great cities of the world: New York, Lisbon, Madrid, Naples, Istanbul, Tashkent, and Beijing. It was the approximate median of the ancient trade routes from China to Europe known as the Silk Road, the location of Mount Ararat, where post-flood life on earth, according to the Bible, began, and the northern end of civilization's beginnings in Mesopotamia. It harbored a rich variety of peoples, governments, climates, religions, and regions through Spain, the Balkans, Greece, Turkey, Armenia, Azerbaijan, Uzbekistan, Kyrgyzstan, China, and Japan. Many laymen, heroes, and conquerors had marched the 41st, among them Marco Polo, Genghis Khan, and Alexander the Great. They had crossed many rivers—the Tajo, the Danube, the Amu Darya—at which I intended to stop and fish.

# The Atlantic

May the tablecloths be dry," my father said when he dropped me off at the bus station in Bridgeport, Connecticut. I had heard the expression from him before, and it sounded to me now as almost a clichéd metaphor for good luck—when the ship is pitching in a storm the steward wets the tablecloths in the dining room to keep the plates and silverware from sliding off the table. As a former merchant marine my father liked to speak about the journey.

It was early December and it had begun to snow.

I was catching the bus to JFK airport, then flying to Savannah to board a freighter to Valencia, Spain. As I stood on the platform, a navy blue overcoat, a slim figure within, caught my attention. Peeking from beneath the hood of the coat was the face of an attractive girl. She followed me to the back of the bus and seated herself across from me, looking somewhat distressed.

"Excuse me," she said, addressing me with a half smile. "Do you know, is it possible to have money wired to the airport? I left my wallet in Guilford down the coast, and there is no time to turn around and get it."

She had saved me from trying to speak to her first. "Where are you going?" I asked her.

"To France, where I live," she said. Her cheeks were flushed like ripe persimmons. "And you?"

"I'm going fishing along the latitude of my home."

"Oh," she replied. "I like to fish. I feel about fishing the way I feel about Turgenev, and all things I know about the country."

The bus began to move.

"Why were you in Guilford?" I asked.

"My father lives there. I grew up in Guilford, but my mother is French. I moved back to Normandy with her when they divorced."

"I don't know your name," I said, "I'm James."

"Yannid," she responded. "But where are you headed?"

"To Spain, then Italy, where I plan to spend Christmas with a friend. I'm coming to Paris in the New Year to meet a fisherman."

"Rouen is only one hour to the northwest by train. Don't stay in some cheap hotel room, stay with me."

"Be careful," I said, "I may take you up on it."

"It's not an idle offer."

She took some paper and a pen out of a small brown leather bag and flattened her blue coat to make a platform for writing.

Yannid Browne, she wrote, 23, Eau de Robec, Rouen.

"I'm a student of medicine at the university in Rouen," she said.

The bus labored through New York traffic and at last stopped at the airport terminal. Yannid stood up and got off, wishing me a good trip, and I wished her one.

My plane from New York landed on time in Savannah, Georgia. I spent the afternoon walking along the Savannah River looking for fish, watching the pearlescent currents swirl. It smelled of both the city and the sea.

I boarded my ship at Garden City Terminal the next morning, accompanied by the port manager, Michael Tomlin. We walked together up the tall stairs of the container ship to its deck. It was there I met Ulrich Günther, the captain.

Dressed in his heavily starched white shirt with bumblebee epaulets, Günther sat us down in the ship's conference room and reviewed my papers. He and Tomlin spoke informally between business matters.

"Join us in the mess at seventeen hundred for dinner, Michael, will you?" Captain Günther said.

"No, thanks, sir, I'm going home to spend time with my wife."

"I wish I was with my wife," Günther said and took his black mustache between his forefinger and thumb.

"We're puttin' up the Christmas tree," Michael said, licking a chubby finger. His face looked fresh and tubbish beside the lean and weathered Günther.

"Early, isn't it?" Günther asked, "or is it a fake?"

"It's a fake one."

"Oh," said Günther, looking down at the papers. "Last year we bought our Christmas tree in Portland, Oregon; this year it will be in La Spezia, Italy. We will spend the holiday in the United Arab Emirates, which I don't like very much."

At dinner, as the only passenger on the 750-foot freighter ship, I was instructed to sit beside the captain. We could hardly hear ourselves over the din of machinery working to unload large steel containers filled with cargo. Loud metallic booms and bangs echoed throughout the ship. Sometimes the cranes screeched under the weight of a container and bellowed like a whale song; sometimes the raucous bangs sounded like a Dumpster full of raccoons.

At breakfast the next morning I met the ship steward, Kokoria. He introduced himself when he came to serve me my eggs and ham. He put down my food and orange juice on the checkered tablecloth and extended his chubby hand. "I am Kokoria," he said.

That afternoon, under the guise of delivering tea and butter cookies, Kokoria came into my small cabin without knocking.

"I am from Maiana Island," he told me, putting down the tea on the table where I was reading. "It is part of the Kiribati chain in the South Pacific." I found the purple aloha shirt he wore amusing. He licked his fingers like he had just been eating a cinnamon donut—his balding head with kinky strands of hair looked rather like a coconut. I put down my book and laughed a bit.

"You are happy, sir?" he said. "Where are *you* from?"

"I'm from New York," I replied, assuming he would not know where Connecticut was.

"I have butter cookies for you too," he said, looking at me. He took an interest in what I had been reading. "Good book, sir?"

"It's the *Odyssey.* You heard of it?"

"No, sir, I cannot read." Kokoria sighed. "But I am learning. What is it you do, sir?"

"I write books and illustrate them, mostly about fishing."

"Oh, good sir, that is good work." He sat down on my bed and took out a pack of cigarettes. "Do you mind, sir?"

"Not at all."

He lit his cigarette and puffed at it delicately between his indelicate lips.

"You should come to Kiribati and write about my island," he said. "But there is not much to do. We only fish and catch lobsters and gather coconuts. You may find that boring. We make wine from the coconut flowers, we catch land crabs to eat. We get our fresh water from a well. We have no electricity, only the sun—equal days and nights all year. Twelve hours of the day we fish, twelve hours of night we, ha ha ha." Kokoria laughed. "There wouldn't be much for you to write about. The women only like it in one position, and if you do it any other way they slap you in the face."

The day grew dark as Kokoria told me stories about life in Micronesia. "You can go down to dinner if you'd like," he said. "No one else will be there, though. The ship is preparing for departure." I stood up to look out the window. We had not yet moved. Detritus edged down the river with the ebbing tide and a yellow moon rose over a wide field.

Finally a horn sounded throughout the ship. "Oh, I must go, sir," Kokoria said, "we are departing."

At half past ten, the moon disappeared from my window as the ship turned south to face out of the channel. I climbed the metal steps to the bridge, where all the ship controls were. A crew member stood at the helm of the ship, directed down the channel by the pilot. Captain Günther handed me a pair of binoculars as we passed historic River Street in downtown Savannah. "It is Savannah tradition to watch the windows in the Marriott for women as we pass to the sea," he said.

Through the binoculars I watched the quaint street decorated for Christmas. Then, as the ship moved, the festive street with restaurants and shops gave way to an abandoned train yard. In the dense scrub that grew there, a fire burned. The moonlight sank and seemed to melt over the breakwater and the lapping waves.

Five hours after we left Savannah, we were pushing through a dark sea under a black sky. Standing astern with my hands on the cold railing, feeling the dried salt on steel, I watched the boat's turquoise legacy. We were sailing within the current of the Gulf Stream. The Homeric Greeks had been right to think of the ocean as a giant river surrounding the world.

Captain Günther was always last to leave the mess at meals. After eating, he reclined and enjoyed several cups of tea with spoonfuls of honey. I sat with him. We were now at sea.

"A beer, maybe, in my cabin?" he said, standing up. "How about nineteen thirty?"

"Sure," I agreed.

Günther kept a good stock of Rostocker pilsner, brewed in his hometown on the Baltic Sea.

"I hope you like it," he said, pouring some in my glass. "I don't know if it's good. I drink it because it reminds me of home and my wife."

"My father was a merchant marine," I told him.

"Then he knows. I'm sure he has many stories, as I do." Günther took a drink of his beer from the glass and reclined in his seat. "So you understand, too, what the sailor's life is like."

We finished our beers.

"Good night," Günther said.

Every day, the ship sailed farther north and farther east, and every day the days were shorter. The wind blew thirty knots at our stern, faster than our speed, which caused the ship to pitch. Lying in bed in my cabin reading, I paused to stare at the blank ceiling. It could have been night or morning, light or dark, I would not have known were it not for my human clock, Kokoria.

The sounds of the crashing waves and constant hum of the ship's engine were soporific. The ocean, besides being a river, was a watery desert, with waves for dunes.

"This is the North Atlantic in winter," Günther told me at our next meal. "Twenty-foot swells. We are nearly at your parallel, thirty-nine degrees north now."

The seas calmed a little the next day, which happened to be Sunday. The captain invited me back to his cabin for afternoon tea, this time with the first mate and chief engineer.

As I entered Günther's cabin I heard a familiar tune. It was "Oh, Christmas Tree," playing in German, as I had first heard it in the house of my Czech grandfather in New Rochelle: *O Tannenbaum! O Tannenbaum!*

Günther handed me a cup and saucer and looked at my face, which I think must have looked pale.

"*Ja,* so," he asserted, "we are getting a little stormy weather. Please sit. You forget it is Christmastime out here on the ocean. That is why we have the ritual of tea and cake."

The diminutive flames of three candles burned beside the plates and the warm sweet cake.

"The Russian cook has made this cake for us," Günther said. "I suppose we should be thankful even though I have to teach him how to cook German food." He cut the cake and gave us each a slice, which we tasted.

"It's not bad," the chief engineer mumbled.

"No," Günther said. "I am surprised."

Several clay gnomes were arranged around the coffee cups. Each gnome had a small pipe, and when the incense beneath their caps was lit, smoke escaped in aromatic tendrils. The three men and I watched the powerful storm begin to wane outside of the window.

The next day at lunch, the first mate, white haired, sallow faced, and overweight, was slathering his customary quarter inch of butter onto a piece of bread when his hand unsuccessfully groped for the mustard.

"Mr. Kokoria," he grumbled like a cave beast. "Mr. Kokoria! *Der Senf*—the mustard!"

Captain Günther looked across the table at the first mate. "Just don't call him Mr. Coconut," he said, "he doesn't like that."

The German officers burst into uproarious laughter, dropping their forks on the floor and pounding their fists on the table. They tried to hold their smirks when Kokoria entered the mess with the mustard.

"Kheem," Captain Günther cleared his throat. "We can proceed to make our ham sandwiches now."

That night on our circle route, we grazed the 41st parallel and headed on a course almost due east toward the Azores.

The next morning Kokoria came to my room to tell me more stories about life in Micronesia. He leaned forward with a childlike glimmer in his eyes.

"I miss the heat on the equator," he lamented. "The crabs, the coconut-flower wine, and the fishing. We have big red snapper. I like to sit in the shade of a palm tree and listen to this old man I know. He knows the Kiribati the way they used to be.

"When a man died, all the people who loved him prepared him for a party. They wrapped him in a blanket of woven palm leaves and then left him in the home until the heat got to him and the water started to come out."

"The water?" I said, putting down my book and leaning forward. "Out of the body?"

"Yes, water starts to come out of the body, just a week or so after death. And when the water comes, the people in the village take him outside and invite all his loved ones to a party. They bring bowls of a mashed root called *papay* and kneel beside him, all helping to unwrap the body. Then, taking scoops of the *papay* in their hands, they sop up the water from the dead man's body and eat it."

"Eat it!" I exclaimed.

"Yes, horrible, I think," said Kokoria. He laughed and slapped his chubby knee.

Nine days later I was standing on the deck with my hand on the railing, a stiff and mild southeasterly breeze blowing at my neck. I heard the captain's voice behind me.

"Land," he said, almost at a whisper, "do you see it through the haze? It's the coast of Spain."

Soon I saw other ships and the ferry to Tangier. As we came closer to shore I saw windmills. We passed through the strait above Morocco and sailed up the east coast of Spain.

I gave my copy of the *Odyssey* to Mr. Coconut, and when he asked what it was about, I told him, "It's about a guy who's trying to get home."

## Spain

Ten hours after the ship docked at Valencia, I was watching the sun set over the Moorish quarter of Granada. I had driven there in a rental car and settled in a hotel on avenida Fuentenueva.

*"¿Es pescador usted?"* I heard the waiter say when I had seated myself at a restaurant in town.

"Yes, I like to fish," I uttered. I assumed he was talking to *me* because there were no other patrons. "How do you know?"

"I know a fisherman when I spot him," he said. A Spanish fishing magazine was showing from my shoulder bag.

"Is the fishing good in Andalusia?"

"The best way to find out is to ask a local fisherman, but not just any, you have to find the right one."

After dinner, I walked a steep and winding street through town. Small white lights glittered from the tall windows of apartments, and although the shops buzzed with people buying presents, I felt a quiet silence in the cool night. I imagined myself leaving the city the next day, taking to the road as the sun was rising and hiking high mountain tributaries of the Guadalquivir in search of trout.

The next morning I visited the *agencia de medio ambiente* in town to get a fishing license. I had been told by a man who ran a fishing store in town to go to the second floor of the building and meet with a uniformed official named Jorge.

"It takes two weeks to process a fishing license," Jorge said. "How long is your stay in Granada?"

"I'm leaving in a week," I said, "does that mean I can't fish?"

He put his elbows on his desk and scratched his head. "*Pues, vale,* maybe I can make you a temporary one for your time here. But I'm afraid you'll have to pay the price for a full year."

"That's no problem," I said. "I just want to fish for trout."

"Oh," he said. "I'm afraid that trout fishing is closed in December. The season reopens in March. This license is good only in the region of Andalusia. It is no good, therefore, in Asturias, Guadalajara, País Vasco, or Galicia, understand? There is only one stream open for trout fishing this season. It is called Rio Frio and it flows through the town of Rio Frio. It is a *coto intensivo de pesca,* which means there is a warden and special regulations." Jorge took a breath and scratched his head through thick black hair. "Do you still want the license?"

"Yes," I said, "*claro.*"

The official stamped my license *caja rural de Granada,* and handed it to me. I paid him the necessary pesetas.

Two hours later I was in the small town of Rio Frio through which the Rio Frio flowed. I touched the water. It was indeed a cold river.

Old men and women, bent over a railing by a bridge, threw balls of bread to the trout. I talked with some fly fishermen who were having a bite to eat on a bench, their rods leaning up against a nearby tree. A young boy displayed his father's catch to me, carried in a reed basket.

"*Truchas arco-iris,*" the boy declared proudly, pointing to the fish. They were beautiful trout, but not what I was expecting. The rainbow trout was introduced from America.

"I hope to catch the native trout," I said.

"Oh, *trucha común,*" said the father. "I don't know if there are any left."

"What do I need to do to get a permit to fish here?" I asked the fisherman.

"Talk to the warden," he said. "He's the man in the green uniform. His name is Pépé."

Pépé was in the warden's shack, his belly spilling over his belt. He wiped the underside of his big round nose with his hand, disturbing a neatly combed Dalí-style mustache.

"What can I do for you?" he said.

"I'd like to fish Rio Frio."

The little wooden shack had corkboards full of photos of fishermen holding big rainbow trout. "You need to make a reservation," he said, licking his finger with his tongue and flipping some papers. "Only fifteen people can fish the stream per day. As you can see, Friday is a popular day," he added, looking at me as if I didn't understand. "Everyone is fishing for Friday dinner."

He looked into a little book of appointments. "Tomorrow we have only six anglers. Would you like to fish tomorrow?"

"I want to fish Sunday."

"Sunday is just four days from Christmas, everyone is shopping, no one else is scheduled for that day."

"I'd like to fish Sunday," I repeated.

"Very well," Pépé said. "I'll meet you here in the guardhouse Sunday morning at nine. You won't want to fish much earlier than that; it's too cold before the sun comes up and the trout won't be biting."

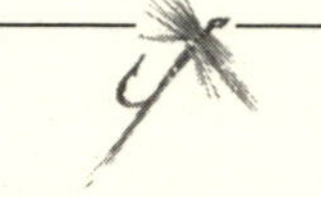

## ALHAMBRA

I had come in part to Granada to see the Moorish palace of the Alhambra. Its many fountains, bubbling springs, and reflection pools amid the orange trees and myrtle hedges harnessed and expressed without words the beauty and language of water.

Toward the middle of the thirteenth century, and just after his return from the siege of Seville, the Moorish sultan of Granada, known as Alhamar, commenced building the splendid palace of the Alhambra. By the early nineteenth century it was in ruins, until Washington Irving arrived in Granada and unveiled its secrets and histories with his prose. I read Irving's tales of love between princes and princesses and anecdotes of daily life in the palace while I was sitting in the palace itself.

"I had repeatedly observed," Irving wrote, "a long lean fellow perched on the top of one of the towers, maneuvering two or three fishing-rods. . . . It seems that the pure and airy situation of this fortress has rendered it . . . a prolific breeding-place for swallows and martels, who sport about its towers in myriads. . . . To entrap these birds with hooks baited with flies is one of the favorite amusements of the ragged 'sons of the Alhambra,' who . . . have thus invented the art of angling in the sky."

I had walked there from my hotel and entered the palace through airy arches and between tall slender cypress. It was a labyrinth of passageways and courtyards that confused its visitor into a fiction. All about, too, was the sound of water as it flowed through chutes between trimmed hedges and fragrant orange groves. I understood that as a people of arid climates the Moors coveted water as others did jewels, and that, as Muslims, water was the synthesis of all things pure. The Koran reserved the parable of water for describing paradise—a place with gardens of flowers and fruits where it flowed without end. Throughout the palace, water was enshrined by the architecture, which forced it to make all manner of bubbling and gurgling music as it slid through long pools and small channels cut in stone.

I stopped to rest on a bench and watched the goldfish in the pools. They gathered and glinted near the surface in the low rays of sun. I walked deeper into the maze of fruits, flowers, fragrances of citrus, green arbors, myrtle hedges, delicate air, and rushing waters. I explored the half-lit tunnels and passageways, the courts and terraces, and towers high over Granada. Water ran in cascades down the steps of a stairway and funneled through the handrails. Everywhere water was spurting and gushing, pleading from dark cavernous corners; sun-bedecked, algae-strewn, reflecting the blue sky, the orange trees, and the white palace walls. The cry of water was heard throughout, and all of Andalusia. It was heard beneath the wheat, between orange and olive trees, in the lachrymose song of the Gypsy, the strings of their guitars, and the words of their native poet Federico García Lorca in his poem "*La Guitarra*":

*Llora monotona*
*como llora el agua,*
*como llora el viento*
*sobre la nevada.*

Sunday morning, I was on the bank of Rio Frio again, blowing on my hands to warm them so I could string the line through the guides of my fly rod. Above the river were orchard hills covered with olives and small stone houses with red-tile roofs. The olive foliage was a soft muted green as if the leaves had been lightly dusted with flour and the rows of them appeared soft like strings of clouds.

"It's too cold," Pépé said, "for me, anyway." He started to walk away. "I'll leave you to your fishing, *buena suerte.*"

The sun had just risen a finger's width above the orchard hill and was not yet strong. As it rose higher it would warm the air and the water; insects would hatch from the stream and trout and swallows would eat them. I headed upstream on foot to wait for that synthesis of events.

The river split into two smaller streams. I chose to take the right fork because it looked neglected. Bushes had grown over it and grasses grew wild on the banks. I pushed through the forest of dense scrub until it cleared, as the stream meandered through an olive grove. I stopped there and stared into a large dark pool for some time until I saw another man's reflection in it.

It was an old man on the opposite bank walking his goats to the river. After they had drunk their fill, he led them up a hill through the olives.

I walked upstream, beyond the goatherd, under an oak forest, my feet crunching dried leaves as I went. The pools in the stream became even deeper and darker and the trees grew thicker and closer to the bank. In one pool I saw several trout making neat swirls as they rose to take emerging mayflies.

I rigged up my fly rod, tying a small dry fly to the line. Then, taking care not to hook the branch of one of the scrubby oaks, I cast my fly into the pool. It landed softly and floated over where the trout were feeding. One took it, gently and swiftly. When it had tired, I steered it to my hand, and held it. It was not the native trout, but a rainbow. I knocked it on the head with a stone and put it in a pocket of my vest. Then I walked farther upstream, through fields of overturned sandy soil.

The pools became clearer and shallower upstream and the olive trees on the banks grew in girth, twisted and pitted with ghoulish craters. The message in Spanish that fishing was prohibited was printed on signs nailed to several trees. Perhaps up here, I thought, far above the village, there were native trout. But as I continued upstream, the signs became more numerous and there was a small house up ahead. I began to feel uncomfortable and after walking several more yards decided to return to the guardhouse.

"How did you fare?" Pépé said when I returned.

"I got one."

"Oh, a good rainbow trout," he said, peering in my vest pocket. "Your first trout in Spain?"

"Yes," I affirmed, though I was thinking that I wished I'd caught a native.

I brought my American trout to the restaurant where I had gone two nights before and they agreed to cook it for me. I asked for the waiter-fisherman but he was not there. The trout was prepared *con vino y romero,* with wine and rosemary, and I devoured it. Then I returned to my room to read.

*Don't sound desperate when you call the French girl,* my inner voice said, *though you are in need of some company.*

I got out of bed, walked to the telephone, and dialed her number.

*"Allo,"* darted a crackly voice at the other end of the line. *"Allo,"* she said again, but she sounded cool and foreign like the drafty hallway. I feared she had returned to her life as a medical student and forgotten the young man on the airport bus.

"This is James," I finally replied. "I met you a couple of weeks ago on the bus. We sat together."

"Of course," she said. It was Yannid again, the blue overcoat, the persimmon-colored cheeks. "I assume you arrived in Spain?"

"Yes, I'm calling you from Granada."

"I wish I were in Spain. It's cold here. I have long shifts in the hospital and exams coming up, but that's more than you want to know. How are you? Honestly?"

"I was thinking about you while fishing today," I told her. "That I'd like to come visit you after Christmas. I'd like to see you when I come to Paris."

"I'm not sure if I'll be here when you arrive. Mom likes to spend the holidays in Belgium with my sister and me. I don't like to go but I probably will. We return shortly after the New Year. And you, where are you spending Christmas?"

"I'm going to Italy; my friend has a house in Tuscany."

"Oh, yes, you told me." Yannid paused for some moments. I sensed that she was thinking. "I told you, you know"— she paused again—"you don't have to stay in a dingy hotel when you come to Paris. I have room in my apartment in Rouen, you just have to wait until I get back." She paused again, this time for longer. "And I hope you don't expect anything."

"Expect anything." I gulped, my heart racing. "No."

"Because it's kind of radical to come all the way from Italy and not expect anything."

"I'm not sure what you mean," I said, and laughed. "But if I do, then I can say that I don't expect anything, but that I do like you. I should tell you that."

"Good," she said with a more official voice. "Even if I go to Belgium with Mom, I will be back in Rouen by the third of January. Call me when you get to Paris and we'll arrange to meet up on the first weekend. I'll take you for a drive in Normandy and then you can settle in. Okay? Good-bye for now. Oh, and merry Christmas."

## TUSCANY

Before I'd left on my trip, my friend Larry Ashmead asked if I'd like to spend Christmas with him and his friend Walter at their home in Tuscany. He could take me eel fishing. I flew to Rome on Christmas Eve and drove north in a rental car, along seemingly endless hedges of oleander, to a wide-open rural countryside.

Larry's home was an old farmhouse in the village of Cosona. It was like others on the nearby Tuscan hills, a stone building on a terraced hill ringed by cypress and olive trees.

Larry greeted me at the door. It was the first time I had seen his warm, inviting look outside of New York. He walked me through his fifteenth-century farmhouse over wide rugs in somber earth tones. He introduced me to his friend Walter, a retired artist and entrepreneur. They had arrived from America the night before.

We sat in couches by the fireplace to rest, eating hunks of parmesan cheese and drinking a local red wine. Olive wood burned a blue flame and warmed the hearth.

"You should know," Walter told me, "that when Larry and I bought the house five years ago, we shined a light down into the cistern and saw goldfish swimming down there. That's the only fish I've seen in Tuscany since we started coming here, but I'm no authority."

"Well, if you can't catch the goldfish," Larry suggested, "we can go visit the eel pots down at Lake Trasimeno; the Italians are crazy about eels."

I took a walk in the early evening down the side of the hill below the house. The tall cypress were almost black against a platinum sky, like sleek dark fishes pursuing the stars. The distant hills were dry, and still bathed in a hint of roseate light from the setting sun.

The next morning, Christmas morning, was bright and clear and the distant hills were a soft violet color. Larry, Walter, and I piled into a Fiat and drove to Lake Trasimeno to go fishing. There was a pier on the lake from which we could see where the fishermen had set their eel pots. The traps were marked by tall sticks that protruded from the milky blue water.

"The eel is a traveler," Larry proclaimed. "You know they spawn in the Sargasso Sea and the young come up the freshwater rivers, even to this little lake in Tuscany, just like they do up the Hudson in New York."

I was casting a small minnowlike lure into the lake when a man driving by on the road stopped his old Land Rover and joined us by slow steps at the lake's edge. He was wearing a jacket and tie.

"You're casting in a good spot," he said, "most of the lake is shallow but here it is *multo profundo.* Let the lure sink and bring it in slowly and maybe you will catch a pike. *Piano, piano,"* he repeated softly. I reeled in my line and watched the lure flash and flutter in the water. Then I cast out again, cranking the reel handle more slowly.

*"Piano."* He pushed his hands down in the air as if he were testing the softness of a pillow.

On the morning of St. Stephen's Day, the day following Christmas, I took an early walk to the village of Santa Anna and a small church there locally famous for its frescoes. The rolling hills were quiet and blanketed by a light frost. Were it not for the sounds of distant gunshots, I would have thought Tuscany was still asleep. The reports were so distinct it seemed as though you could follow them with your eyes across the smooth hills.

Then I saw a hunter crawling out of an old Land Rover, slipping with his dog into a pocket of brush in the open countryside. He wore a green waxed jacket, wool pants, and leather boots. Others plodded across the hillside shouldering their guns, making straight tracks through the soft, newly tilled clayish soil, while their spaniels, limber and amber eyed, crisscrossed the countryside intent on the smells of game in the still air.

Toward midmorning the frost was all but gone, lingering only in the shadows of houses and barns. Finding that the church with the frescoes was closed, I returned to the house for lunch.

Larry and Walter had prepared a big ham, which they had carried from New York. The smell of the baking pig filled the brick-arched interior of the old house. We ate more *pici* with melted pecorino cheese and swilled more Tuscan wine with an arugula salad sprinkled with black truffle oil.

Larry said, knowing I was leaving the next morning for Paris, "Your last chance for a fish in Tuscany is the goldfish in the cistern. Shall we try for him?"

I explained to Larry that it wasn't the kind of fishing I normally did, but it was a worthy exercise and I might as well try. He had found the key to the cistern top and, before I had my rod rigged with a hook and a small bit of ham, he had unlocked it and was staring into the dark black hole.

"Try a bigger piece," Larry suggested, watching me bait my hook. "I remember it being a pretty big goldfish." I lowered the line into the water and left the rod while we went inside to warm up by the fire.

We sat by the fire for a while and I talked about where I planned to go when the weather improved the next summer. Larry offered no opinions or suggestions; he just listened.

"Let's go see how the line is," he began, after a while, so we went out to the cistern. I picked up my rod to reel in the line and, to my surprise, something was pulling at the other end. I lifted the tip of the rod and watched it tick as the fish swam in tight circles. I put the rod down and took the line in my hand, pulling a sleek black fish from the water. It was an eel.

As I looked down into the cistern, there was enough light from a lamp on the house for me to see my reflection in the bottom. It was like looking into the round hole of the Pantheon's ceiling and seeing your reflection in the sky. I asked Larry how he thought an eel had ever got in there. "I don't know," he said. "I'm just your editor."

The next day when I left, he gave me a book of poems by Eugenio Montale.

"There's a famous one about eels," he said.

## WHAT IS A FISH?

My father's heroes were seduced into learning by a curiosity about the natural world, and he described the ongoing process of that seduction as a person's *loucura* (Portuguese for craziness). Darwin's was beetles, Nabokov's butterflies, Audubon's birds. As a child in Brazil my father fell in love with birds, and birding continued to be his raison d'être in New York, where he grew up, married, and had children, and in Connecticut, where he divorced.

You are lucky to have one too, my father said. Yours is fish.

When in its element, the fish is seen by us in glimpses—an impression of fins and scales gliding over coral, a flash of silver and blue emerging from the depths of the ocean, materializing from the gravel at the bottom of a mountain brook.

Fish are also the creatures, some say, from which we have evolved. The limbed lungfish has been left along the evolutionary line as some clue of our connection to life in a primordial soup. We ourselves are nurtured, fishlike, in fluid before we breathe air, and are constructed largely of the medium within which fish play.

Fish are both the object of human fantasy—the mermaid, half golden locks and breasts, half scales and tail—and the representation of the sublime and horrifying, the white leviathan of Ahab's monomania. For Far Eastern peoples, the fish is the symbol of peace and order, strength and perseverance. It inhabits Japanese Zen garden pools. It is the meal of good fortune on the Chinese New Year, the auspicious symbol of boys' day in Japan, the representation of the eye for Tibetan Buddhists. The fish is holy for Jews on Passover, the early symbol of Christianity, the miracle of Jesus, the food that bestows immortal life, the Friday dinner. It is Pisces, the twelfth sign of the zodiac, which denotes the end of the astrological year and also the beginning.

The dual aspects of the fish as both a monster-seducer and a symbol of goodness and immortality have fascinated psychoanalytic minds. Fishes, Carl Jung writes, are said to be "ambitious, libidinous, voracious, avaricious, lascivious—in short, an emblem of the vanity of the world and of earthly pleasures," but they are also a symbol of "Messianic significance," and of "devoutness to religion."

In a Boeotian vase painting the "lady of the beasts" is shown with a fish between her legs, or in her body. In drawings of the Japanese artist Hokusai, an octopus makes love to a woman. In ancient Chinese erotica, the penis entering the vagina is sometimes described as a loach (a small bottom-feeding fish) burrowing in the mud. A passage in Günter Grass's novel *The Tin Drum* describes a woman taking pleasure with a live eel. In ancient Rome a mullet was sometimes inserted into a woman's vagina as punishment for adultery.

In fifth-century China, flatfishes, which have both eyes on the same side of the body, were thought to represent both the male and female, the male on the left side, the female on the right. Consequently, in order to see in both directions, these fishes had to swim together always, side by side. Thus, they became a symbol of marital love.

We cannot account for why exactly it is that we are drawn to fish; perhaps it is as Hermann Hesse says in his novel *Narcissus and Goldmund:*

For a while Goldmund sat on the embankment. Dark, shadow-like fish still glided by down there in the crystal greenness, or were motionless, their noses turned against the current. A feeble gold shimmer still blinked here and there from the twilight of the depths that promised so much and encouraged dreaming.

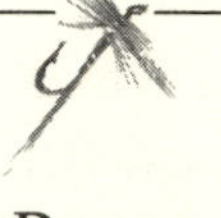

## Paris

My train to Paris left from Florence. It passed through the mountains in the dark and, come morning, along the blue-green Côte d'Azur where palm trees grew. I saw men in a boat dropping nets and men with fishing rods casting from rocky breakwaters.

I was thinking about Yannid and meeting an extraordinary fisherman I had heard of but not met, Pierre Affre. Whether I would spend the winter alone or with friends depended on my success in cultivating relationships with these two.

I arrived at the Gare de Lyon and settled into a hotel on rue de Caumartin not far from l'Opera. Then I phoned Yannid at her apartment in Rouen.

"*Allo?*" she greeted in her soft voice.

"Yannid, it's James." I hoped she would remember her promise, our planned meeting. "I didn't expect to get you," I said. "How are you? I'm in Paris."

"I'm fine. You're already here?"

"You're not going to Belgium with your mom?"

"I was too busy to go," she explained. "I'm not going to Paris for the New Year either, I've got to stay here and study. I'm sorry if you planned on my joining you."

"No, it's no big deal." I felt my words were awkward. "I'm in a hotel. I'd like to see you."

"I agree that New Year's is no big deal."

"But maybe I can see you later this week."

"I was thinking we could go to an art show at the Grand Palais. I've been wanting to see it, of Chinese art. I could come in tomorrow and you could meet me there."

"Why don't you meet me at my hotel," I suggested. "I'm on rue Caumartin, near the Opera."

"Okay, that's fine. I suppose I need a break from work."

Yannid arrived by train at Gare Saint-Lazare from Rouen at noon the next day. I was waiting for her in the hotel lobby when she walked in from the cold and damp day.

She held herself with an air of propriety and carried a backpack that looked to be heavy with books. She insisted on taking it with us as we walked to the Grand Palais.

"I come from one of those old Norman families you read about in Flaubert," she told me as we sauntered along the quai. "My mother lives in a big apartment with a view of the cathedral; it's beautiful." She looked at me and seemed to acknowledge an attraction between us. "You have to come see it." We were getting along, just walking through Paris together, getting lost and somehow always ending up by the river again, which looked high and powerful. She liked to talk about art and books.

It was a cool Saturday morning when, several days later, I made the trip by train to Rouen to see her. The medieval city was also on the river Seine but downstream of Paris and closer to the sea. Yannid picked me up at the station in her mother's Citroën. I was shocked by the paleness and luminosity of her face; she looked as if she hadn't slept for days, but I found her slight and sickly appearance strangely beautiful.

"I feel," she told me, "that Normandy is like Connecticut. You'll see the resemblance when we get out into the countryside."

Yannid had told her mother that I would be staying with her in her small apartment. "This is a big deal," she did not fail to tell me, "my mother is very . . . Catholic. She seemed okay with it, though, which kind of scares me. She's given me freedom to do what I want. I guess I'm an adult now, but I find that a bit fearful."

We drove to the little street where Yannid lived, rue Eau de Robec. In medieval times Eau de Robec was a drapery and linen manufacturing center and the little stream that ran through it, the *eau,* or water, of Robec, was said to change color with the dyes that were discarded in it (this is described in the opening of *Madame Bovary*). The water, fed by a spring, now ran clear over pavement and pebbles, and the street was no longer lined with linen manufacturers but with sex shops and cafés.

Yannid stopped in the narrow stone street and opened a nondescript metal door. We walked into a dark corridor and up five flights of narrow spiral stairs. Her apartment was in an eighteenth-century building, and was in a kind of attic or aerie—there was a kitchen, bathroom, and a room with a bed, a couch, and two desks. The room's untidy appearance did not distract me from seeing its crowning virtue, a view of the large rose window of a gorgeous church, gigantic as the moon appears when it is near the horizon. The church's name was Saint-Ouen and it seemed so close you could put your hand through the rose window as though it were a pool of water.

"I couldn't live without this view." Yannid nonchalantly opened the windows. "I couldn't study without it. I mean that, it's my sanity."

I looked out over the sun-drenched slate-colored rooftops to the church and its rose window that was mesmerizing me. Now with the light fading behind passing clouds it looked more like a purple pool of water, almost, with fishes leaping out of it. "I can't wait to

*Goldfish in reflection pools of the Alhambra, Granada, Spain.*

JAMES PROSEK

get out of here. Put your stuff down," she demanded, "and let's go to the beach. I love the water."

"You won't believe how beautiful Normandy is in spring," said Yannid as we drove up the hills outside Rouen. "Now I think it looks like Connecticut in December. It seldom snows here, though." In the leafless treetops were balls of green foliage, an evergreen called *gui,* or mistletoe. "My father used to say, under every Yankee's skin there's a Norman. I think he meant that to mean that my mother is cheap. The irony is that he's the tight one."

Yannid and I had lunch at a restaurant by the sea. She had a militant and proprietary way of ordering food, insisting that I get the fish that was in season. So I had steamed mussels and a bland fish soup and was happy to be spoken for. We had a little white wine too, which was good for making conversation and warming the fingertips, and after lunch we walked barefoot on the beach. In the distance was a train of horses and riders, trotting along the edge of a light surf.

We crossed little streams of salt water, which rushed and coursed through the sand, and Yannid told me that the French word for channel or rivulet, *rigole,* came from the verb *rigoler,* "to laugh." "It's a nice way to describe the sound of water, isn't it?" she said.

"I like being with you," I said to her. "Maybe I can take you to fish for trout. Maybe you could come with me."

"I'd love to, James," she said, "only, close by. As much as I'd like to, I can't travel the world with you, I'm a medical student."

Later that afternoon we sat drinking local cider in a dark pub in a small harbor town lined with sailboats. The aroma of apples brought me home, as did the smells of wood burning in a fireplace. On the way back to Rouen we stopped at a wine shop and Yannid bought two bottles of smoky cider and some apple brandy. It was just nice to be with her, to drink and be lost in a place I didn't really know. I wanted to go where the going took me.

* * *

When we returned to Rouen in the evening we stopped at the apartment where Yannid's mother lived in order to drop off the keys to the car. "Mom's undressed for the night and won't be seen," Yannid said, "but come inside because I want to show you something."

It was a two-floor apartment furnished with beautiful hardwood chairs, chests, and tables. Yannid led me to a window beyond a tall armoire. "It has the best view of the cathedral of any apartment in Rouen," Yannid explained, pulling back the blinds to show me.

"This is the cathedral Monet painted in all its changing moods as storm and sunlight swept through town. It's stunning, isn't it? And yet I don't really like Rouen anymore." The cathedral was lit mysteriously by floodlights from the ground and behind it was a dark purple sky. Yannid turned from the cathedral to look at me.

"Are you sure this is what you want?" she asked and kissed me.

"If it's what you do," I answered.

"I can be one of the little fishes you admire so much," she said and put her arms around me.

"I would like to show you my cathedral, the trout stream."

"The way you describe it," Yannid replied, "I want to experience it with you." She loosened her embrace and kissed me on the cheek. "Let's go back," she whispered, and we walked through the cool evening along dark streets to her apartment.

The first thing I had noticed about Normandy was that the days were shorter than at home (I was decidedly north of 41°N), the second was that it never got very cold and that it rained a lot. It seemed to rain for weeks, and the level of the Seine continued to rise. I found it hard to leave Yannid's cozy apartment, though I contemplated later taking out my fishing rod and trying to catch a fish—it was not for lack of time to myself that I didn't; I think I was intimidated. I rarely saw Yannid because she was either in class or on a shift at the hospital. I began taking long walks along the river, watching it swell with more water before my eyes, and every day, until it resembled a sea with waves, and on windy days there were whitecaps. I couldn't even consider how I would fish it. I needed to find a guide.

## Meeting the Retired Veterinarian Pierre Affre

For years before he introduced me to him, my friend Nick Lyons had told me stories about a mad sports-fisherman from Paris named Pierre Affre. As an author, publisher, and traveler, Nick had met many fly fishermen, and all sorts of enthusiasts and aficionados of the rod and line. Pierre, though, through his skill, intuitiveness, dedication, and outright weirdness, was according to Nick above the others, and of nearly mythical status. One of Nick's favorite stories about Pierre was about how he'd managed to get a hook lodged in his penis while tarpon fishing in the Florida Keys.

"Now a tarpon, as you know, is a big fish with a hard mouth," Nick explained as he told it, "so you need a sharp hook and a big hook. I don't know how he did it, but it was a windy day and the line came sweeping by his pants as he cast it, and there it was. I'm in pain every time I think of it."

I later collected many of my own Pierre stories; one in particular that I heard from his close friend Peter best exhibits his true *loucura* for fishing. "I used to accompany Pierre when we were in our twenties to Iceland to fly-fish for salmon in summer. As you know, it's near the Arctic Circle and in July there are no nights. Pierre used to take speed pills so he could fish 'round the clock. He's a mad fisherman; I don't think he slept or ate for five days, he just fished."

Nick shared with Pierre an understanding of the fisherman's (and thereby the predator's) folly. Fishing is a philosophy to them, a lifestyle, a source of frustration, and also of comfort, the same philosophy understood by the man who had introduced them to each other, the late hotelier Charles Ritz.

After months of anticipation, I called Pierre one morning in his office on rue Dauphine in Paris, in the sixth arrondissement. "Ah, James," Pierre said softly yet urgently, "I have been expecting your call. Nick wrote me a very nice letter about you and sent a copy of your book. We have a lot to discuss. It is important for you to come to Paris as soon as possible to see me."

"Okay," I said quickly, for I was very excited too. "I will get a room in town."

"Don't worry about a room," Pierre said more strongly, bringing out his French accent, "you can sleep in my office. I would put you up in my apartment but I have three young children and no space. Anyway, Nick tells me you are with your fiancée in Rouen. Well, it won't take you long to get here. And keep in mind that in two weeks is the big French fly-fishing exposition, Le Salon de la Pêche à la Mouche. I am organizing it now for the fifth year; you will enjoy this immensely. If you come a few days before the *salon* you can help me with the setup. You will meet some fantastic fishermen from all over Europe, good guys, from Russia, Iceland, and Holland especially."

I pictured Pierre speaking from a dark apartment in downtown Paris surrounded by books with the river nearby and the rain falling. I had only communicated with him by letter, so it was a privilege now, and a bit surreal, to ask a question and get an immediate response. "Can we fish the Seine?" I asked.

"It is a bit high now, but of course we will fish the Seine. The better fishing is in May and June, but we have an unusual fishery now off the Ile Saint-Louis. I will tell you all about it when you arrive. We catch big pike and zander and even the occasional sea trout, do you believe it? In downtown Paris. It is an amazing fishery, as you will see."

## The Old Brothel on Rue Dauphine

As it turned out I could not go to Paris immediately because I had planned a trip with Yannid to go skiing in eastern France. But I called Pierre the moment I returned. When he answered the phone, he was breathless with excitement.

"You must come quickly," he urged. "We've been having fantastic fishing in the Seine. I have never seen it like this before. We're catching *silure,* up to thirty kilos. You won't believe it."

*Silure,* I learned, is a catfishlike fish with a flat head, a wide mouth with six whiskerlike barbels, and a long tadpolelike tail, which it uses to stun its prey. The *silure*'s cone-shaped body is designed to eat and digest big things like other fish, small ducks, geese, and garbage on the river bottom. This accounts for its immense size—it is the second largest freshwater fish in Europe after the sturgeon, and specimens have been caught in excess of two hundred pounds (the largest from the Po River in Italy).

What made Paris extraordinary for me was its intimacy with the Seine. At so many points you were drawn to walk beside it, or across it on its diverse bridges, acknowledging it as an artery without which the city's beauty would pale.

Pierre's narrow street, rue Dauphine, was perpendicular to the

Seine, an extension of Pont Neuf on the Left Bank not far from Notre-Dame cathedral. The entrance to rue Dauphine from the quai was one of those magical points in Paris where in just several paces you crossed from the majesty of the riverine city to the narrow street of a seemingly smaller town.

The morning I arrived from Rouen on the train it was drizzling in Paris, and by Pierre's reports it had been raining for days. Though I was not familiar with the Seine's normal flows, when I crossed by foot on the pont Neuf, I could see that it was in flood, so high that barges with tall loads could no longer slip under the bridge arches. The water was opaque and yellowish, a combination, I thought, of suspended silt and reflections of the sandstone buildings on either bank. There was no way to see a fish in it—the visibility could not have been more than two inches.

I walked in a nondescript door at 23, rue Dauphine and up a warped spiral staircase to the third floor, where I approached a red door in a dark corner of the hall. I assumed it was Pierre's because there were six scales from a giant tarpon nailed to it in a circle. I knocked and heard quick footsteps across a creaking wood floor. The door opened and a man of average height stood in the frame.

"Please," he gestured, "enter."

He looked like my mental version of a rural apothecary, blue eyes staring at me through elliptical wire-rimmed glasses resting on the bridge of a cheerful gnomish nose. He wore rubber boots, a plaid shirt, and corduroys, and had two days or more of stubble on his face. His hair was matted on the back of his head from sleep, his sea blue eyes not alert but aware.

When my eyes adjusted to the small space I saw there were two rooms. Against the walls were shelves from the floor to the ceiling choked with thousands of objects: books, magazines, terrariums with taxidermied fishes, fish spears, harpoons, plaster casts of fish, reels, rods, silk lines, and carousels full of slides.

In the second room by a pair of frosted windows was a desk piled with books and paper and an antiquated computer. On top of it all Pierre opened a world atlas and bade me to sit on the opposite side. Flipping pages gently, he perused parts of the 41st parallel and made suggestions on where I should go based on places he had been.

In his forty-five years, Pierre had fished many of the world's major rivers: the Volga, the Nile, the Amazon, the Amur. I pulled my chair closer. "What's your favorite fish?" I asked him.

"Oh, I like them all," he replied, "but if I had one day left to fish I would probably go to Sierra Leone for giant tarpon. My favorite fish though is the Atlantic salmon. In my twenties I was a guide on several salmon rivers in Iceland. I wrote my thesis for veterinary school on the Atlantic salmon, which by that time, in the mid-seventies, was in very poor shape in France. The Allier, a tributary of the Loire, is my favorite salmon river in France, that and the Gave d'Oloron in Basque country. The salmon once ran the Allier by the thousands." He flipped to a map of France. "They congregated here, between Brioude and Chanteuges, to spawn. Now we would be lucky with a run of three hundred. For fifteen years I fished the Allier, an average of twenty days a year, and I caught one salmon." Pierre paused to clean his glasses on his shirt, then refocused his attentions on the map.

"While fishing the Gave d'Oloron I became friends with an old man who fished for salmon commercially with a fly rod. He averaged a hundred salmon a year, and back then, to a poor man, two salmon were equivalent to a month's wages. That old man taught me how to fish for salmon. I loved him."

Pierre talked about native brown trout in the Pamir Mountains of Kyrgyzstan and Afghanistan—"When the shah was still in power in Iran it was possible to drive there from France." He also spoke of a strange trout that lived in spring-fed streams in the Balkans, the giant brown trout of the Aral Sea, the taimen of Mongolia. It was alternately drizzling and raining outside the dirty windows as Pierre spun his tales.

"I landed a huge taimen one night in the Orhon River, or *Gol*, thirty kilos! and tied a rope through its jaw and around a log hoping to take photos when first light came the next morning. I was up with the sun and left the *ger* [a Mongolian tent for which the Russian word is *yurt*] for the river carrying my rod. The big fish was still in the small eddy where I left him the night before—there must have been three feet between his dorsal fin and tail—he was still alive; I intended to photograph and release him. What would I do with such a big fish? I went away to fish a bit and wait for ideal light. Two hours later I returned to the spot and the fish was gone. The ground was covered with hoofprints. Some bloody Mongolian stole it. Probably fed the whole village.

"Well," Pierre concluded, looking at his watch, "there's just too much to talk about all at once; let's return to my place for lunch and we can speak about the fly-fishing *salon.*"

Under a light drizzle I followed Pierre to his home, several doors down from his office, also on rue Dauphine.

"So what is the name of your fiancée?" Pierre asked me casually.

"Her name is Yannid." I didn't find it worthwhile to tell him she was not my fiancée.

"She has a very nice voice," he remarked, looking straight ahead through his clerical glasses, his wavy matted hair becoming wet under the drizzle. "I talked to her on the phone when I called you the other day. I assume you will return to France often to see her—of course—you have to paint all the beautiful French trouts." He laughed. "In the east of France we have a very interesting trout we call *truite zébrée.* This is a very unique fish; it has dark vertical bands on it like a zebra. I will introduce you to Philippe Boisson; he knows the rivers there better than anyone; well, maybe I will go too. The *zébrée* is the trout that Gustave Courbet painted that hangs in the Orsay Museum down the street on quai Conti. You must go see it."

"The house used to be a *bordel,*" Pierre told me as we entered a dark hall over a patterned tile floor and up a small stairs. "As you will see, downstairs there is an exit onto rue Mazet where the men would leave so they would not be seen coming and going from the same door, or the same street."

In a large sunlit room on the top floor of Pierre's apartment, I met Pierre's wife, Carole. She was preparing lunch for two of their three children, who were screaming as they chased each other around the room. Carole looked at Pierre with one hand on her hip and the other on her cigarette. She was tapping her foot, her apron besplattered with macaroni and cheese. Then she turned to the children. She spoke what seemed to be immaculate French, but with an American accent.

"Marlin! Venice! *Regarde-moi!*"

"*Qu'est-ce qu'elle fait?*" cried their father.

"She should be doing her homework, Pierre."

"Do your homework, Venice."

"Papa!"

I remembered when I saw the paintings on the walls that Nick had told me about Carole being a painter with an unusual subject. The one before me was an oil about a meter square depicting a close-up of an animal's genitalia, I know not what kind; tender pink flesh resolved to a wrinkled pile of purplish skin. As an organic abstraction it was beautiful, like an O'Keeffe painting, but not flowers.

Ignoring the chaos about the kitchen, Pierre took me down a narrow stair. Halfway down we encountered his third child, May Fario, a dark-haired girl in adolescent bloom who, like the youngest, Marlin, was named by Pierre for a type of fish (*fario* is the subspecies name of a brown trout native to western Europe).

There were two rooms mysteriously set on half floors off the stairs and then two more on the floor we descended to. The room, unlike the one we'd entered upstairs, was dimly lit. It was wide and fortresslike and the walls were uneven and richly colored, alternating yellow and red. There was only one window, tall and thin in the corner of the room, and a plain metal door that exited onto rue Mazet.

Ragged and cracking skin mounts of pike and trout, some in terrariums with sprigs of lake weed and logs, were hung about the room, musty and dusty like relics of an abandoned mansion. There were also plaster casts of large zander and carp that Pierre had caught in the Seine, and hanging by chains from the ceiling were giant metal sculptures of horrifying insects with chain-saw wings, enormous steel hooks, oil-can collars, antennas of knife blades, bodies of mammal vertebrae. "You will meet the artist who made these tomorrow night when we set up for the fly-fishing show. He's a true original. Would you believe he's a tax inspector and does all this art on the side? You will be astounded by his atelier; it's a wonder he can make any of this stuff there, it's hardly large enough to stand up in."

I thought my eyes had feasted on every nook of the room until I looked at the ceiling, where Carole had painted a Sistinesque rendition of naked cherubim riding dolphins firing gilt arrows at salmon and tarpon with miniature bows. The faces on the cherubim were those of Pierre's fishing cronies, a cast of diverse characters that I would come to know over the next several months.

The more I looked around the room, the more it began to live and breathe. The long, slender, and thick-shafted pike gnashed their teeth from the wall, but with glazed and blind eyes. Two of the iron insects flew about, flapping their chain-saw wings, spreading their shadows on the blood-and-urine-colored walls. Pierre's wild children, seated at lunch upstairs before a painting of an animal's vulva, were a product of this place, this womb-room, this dangerous erotic playground, never certain whether to take the comfort of the couch or to flee the grotesque assemblage of anatomy and art.

When I thought we could go no deeper in the old brothel, Pierre took me down a small stair into what he called the *cave,* a small basement made warped and Gaudi-like by centuries. The cave was his fishing room, thick with feathers and flies, hand-forged fishing spears, and all manner of materials worthy of a creative inferno. Hundreds of fishing rods leaned up against the ancient mortar wall, resembling in a row the baleen of a whale's maw. In this catacomb, a lifetime of angling detritus rested. "So," said Pierre, turning off the light, "I think we should eat a little, and then I'll show you where we fish in the Seine."

## The Ile de la Cité and Le Salon de la Pêche à la Mouche

Papa, Papa!" cried little Marlin, looking up toward his father.

"No, we cannot bring the fishing rods," said Pierre. Pierre hoisted Marlin onto his shoulders and we walked in under the drizzle to Pont Neuf and down to the place du Vert Galant over wet cobbles to the tip of Ile de la Cité.

"There are two islands in the river here in Paris," Pierre explained, "Ile Saint-Louis and here on Ile de la Cité. These are the two best places to fish. When the river is in flood the fish stack up in the eddy downstream of the island where the current is broken. There are probably five hundred bream here. They especially don't like current, and at times the big *silure* come to feed on them. You would not believe, but you will see! They open their giant mouths and eat a four-pound bream whole. While you were skiing, my friends Jean-Pierre and Guy hooked eight off Ile Saint-Louis. They landed one that was about thirty kilos and lost one they said would have been nearly fifty—that's a hundred and ten pounds!"

West of where we stood was the metal walking bridge, the Pont des Arts, and on the Left Bank, the Musée d'Orsay, on the right, the Louvre. Upstream were Ile Saint-Louis and Notre-Dame, whose bells indicated it was well into the afternoon.

"This is the most beautiful pool in the world," Pierre proclaimed, passing Marlin to me as he climbed over a rail, "and the fishing gets better every year. The Seine was very much polluted in the sixties and the only fish that lived here, if any, were the carp and eel. Now there are thirty-eight species in downtown Paris. I only started catching *silure* two years ago, small ones. This is the first year we have caught such monsters."

"And you eat them?" I asked Pierre.

Pierre looked at me through his glasses, raising his gnomish nose and scratching his wavy hair. "Oh, yes, we eat them." He smiled as if to assure me that I would eat them too. "The Seine is clean; in a year or two my friend André insists he will swim in it. The source of the river in Burgundy is a spring-fed trout stream. We will fish it in May and hopefully will find a nice hatch of mayfly. Both François and Vincent have access to excellent trout fishing on the haute Seine."

"Did salmon ever run up the Seine?"

"Yes, of course, all the rivers draining into the Atlantic in France had salmon. The run in the Seine was magnificent. The last one caught in Paris was in 1956." Pierre stopped to carry Marlin up the steps to Pont Neuf again by the statue of Henri IV. "Well, who knows," he continued, "maybe we will get one tomorrow morning. Maybe Guy and Jean-Pierre will join us. We'll fish until about nine in the morning and then we must go and set up for the fly-fishing *salon.*"

That night, I slept wrapped in a wool blanket on the floor of Pierre's office. At eight he knocked on the door, came in with his key, and woke me. "Well, we slept too late," he said, standing over me, "but that's okay, I think. We have a long day ahead." He looked at his watch. "We'll walk down to Ile Saint-Louis anyway, to see if Guy and Jean-Pierre caught anything. We'll leave the rods behind this morning."

Guy and Jean-Pierre lived in a poor neighborhood on the outskirts of the city. A chef in a school cafeteria and a landscape worker in a cemetery, they supplemented their meager incomes by fishing most every weekday morning before work and selling their catches to local restaurants. As far as Pierre knew they were the only commercial fishermen in Paris, and illegal ones at that. Their most coveted catch was the zander—or, as Americans called it, walleye—a sweet, white-fleshed fish, although they also got a good price for eel, *silure,* and carp. Guy and Jean-Pierre could earn up to a third of their normal monthly incomes through their fishing.

Pierre and I walked along the Seine on the quai des Orfèvres to the façade of a newly renovated Notre-Dame cathedral, dim and ominous in the morning light. "They are wonderful fishermen, these guys," Pierre told me of Guy and Jean-Pierre. "I met them four years ago fishing on Ile de la Cité. One of them had caught a ten-kilo zander. I couldn't believe it! So I started to follow them and fish with them and made a small film of their fishing. I gave Guy a cell phone to call me if they caught a big fish, and would ride my bicycle down to the river with my camera. I got them sponsorships through American fishing tackle manufacturers, so they have good free equipment. They mostly fish with worms and they are very skilled. It takes a great deal of skill to fish a worm correctly—as much as it does to fish a wet fly."

At this hour of the morning Paris was quiet and the city belonged to the fishermen. The river was eggshell brown and made the limestone façades of the city buildings seem more luminous yellow and the zinc roofs a deeper gray. "They're out there, see?" Pierre said as we crossed Pont Saint-Louis to the île. The fishermen were two dark shapes on the tip of the island standing with their lines in the water.

"'*Jour,* Pierre," one greeted as we approached them. They had caught a bream and killed it. It lay bloodied on the cobbles. I shook their hands.

"Well"—Pierre looked into the opaque water—"if the bream are here the *silure* should be too," he said. We stayed and watched them for a half hour, after which time neither Guy nor Jean-Pierre had caught a fish, so we left to go set up booths at the fly-fishing exhibition.

The Salon de la Pêche à la Mouche was an annual fly-fishing exposition at the Espace Auteuil near the Bois de Boulogne in west Paris. Represented there were a hundred or more vendors from Austria, Belgium, France, Holland, and Iceland, selling everything from flies and fly rods, wine and swine, to antique salmon spears and trips to exotic destinations.

Pierre was certainly the prince of the show—everyone seemed to know him. He filmed and aired a weekly fishing program on French television and edited a fishing magazine called *Pêches Sportives.* When we got to the space, Carole, his wife, was setting up a booth with some of Pierre's antique tackle and fishing books. I was to help Pierre set up a kind of museum-style exhibit he'd prepared on the history of salmon fishing in France. During the course of the day, we drove several van loads of wares across town from rue Dauphine to the Espace Auteuil, including some sculptures by Pierre's friend François, which would be for sale in a small art gallery space.

As a consequence of our work, I met François Calmejane, the tax inspector–artist who had created the big iron sculptures of flies that hung in Pierre's apartment. He looked a bit like an inspector type with a bowler hat, a thick black mustache, and a Sherlock Holmes–style pipe with a deep curve in it. He was wearing a bright yellow shirt and a tie made of wood, and over that a vest of green ostrich leather. The hair on his head was a kinky dark brown. He held the pipe between his yellowed teeth as he spoke gently and affirmatively. "Just one moment, François," Pierre called, seeing him try to lift a heavy sculpture by himself, "we're almost done; we can help move some things."

"François, you know," Pierre told me in confidence, "is one of the top tax agents in France. He's busted a lot of big guys doing corrupt things. He takes four weeks of vacation every year; two of them he spends with his wife camping and fishing for trout in Ireland, the other two he spends making his art."

I imagined the inspector up in the small attic space above his apartment that was his atelier, creating the visionary sculptures in a two-week orgasmic gesture from ideas and energies stored up for an entire year. At that time I didn't really have a sense of who François was. Though he seemed a little cold, I could say that I had never seen anything like these sculptures he had made. Clearly he had an obsession with fish, and when I had the opportunity to see his studio, I felt I would be walking into his mind, which would be strange but also familiar because I thought I might see a bit of myself in there.

The sculptures themselves, besides being sublime and terrifying, were dangerous to carry, for they had all kinds of sharp objects protruding from them—scythe blades, giant hooks, spears, knives, and chains. You had the sense that if you fell on one you'd be impaled and bleed to death.

All of the sculptures of flies and fish were imaginative and brilliant, but one in particular was, to my mind, a masterpiece. Pierre thought so too, and said so when we had set it up and were standing beside it.

"This one I think is his best work. It really is remarkable. François calls it *le grand bécard vainqueur!* The great male salmon that won. The head"—Pierre pointed—"is the actual head of a salmon that I caught in the Baltic and brought back for François, and the fishing reels in its stomach are mine too."

The sculpture depicted a large salmon with its tail touching its head as if it were leaping in triumph. Its body was a series of curved wires creating a cavity that could be seen into like a cage. Its tail was a fishing-rod handle, and an explosion of various lures hung on spiraling lathe chips. Its pectoral fins were large gaffing hooks. Its air bladder was a gas tank from an old French motorcycle, its intestine was a scythe blade, and in its stomach were the various items of the fisherman's kit—reels of different shapes and colors, lures, flies, and a landing net. They symbolized a salmon that had overcome all the obstacles it faced on its journey up the river from the ocean; it had swallowed all the anglers' tackle, broken their lines, and cursed the industry of man.

Pierre and I were on Ile Saint-Louis the next morning with our lines in the water. Pierre fished with a worm and gave me a heavy rod rigged with a big lure for *silure.*

"I think Jean-Pierre and Guy are fishing at Neuilly," he said. He picked up his cell phone to call Guy to get a report.

"What? Yes, Guy? Two bream, no *silure,* okay. What? *Vraiment?* Okay, *à tout à l'heure."* Pierre put the phone in his pocket, laughing and shaking his head.

"Did they catch anything?"

"They caught two bream and Jean-Pierre hooked an ear."

"An ear?" I asked, thinking I'd heard wrong.

"Yes, apparently," Pierre said, casting.

"You mean a human ear?"

"That's what he said."

"What did they do with it? Are they going to tell anybody?"

"What, and cause trouble? They are just fishermen."

After an hour of fishing we'd caught nothing so we left to be at the exhibition early—it was the first morning that it was open to the public.

Pierre spent the morning trying to sell exotic bird skins to Dutch fly tiers who used the feathers to make salmon flies. "We don't need parrots and blue chatterers," one said.

"How about these bustard feathers," Pierre offered.

*Out of Yannid's window, l' Eglise Saint-Ouen, Rouen, France.*

"They are no good to us without the matching feather from the opposite wing."

"Five thousand francs for all of them then," Pierre said.

"You are asking too much, Pierre," one responded. "Your birds are not in very good shape." They turned the birds over and over in their hands and folded the skins to expose the feathers and show how sparse they were. "I would take all of the skins for a thousand francs and even that I think is a lot."

"One thousand francs?" Pierre gasped, tucking his chin in close to his neck and lifting his shoulders, trying to sound insulted. "They're worth more than that; I could get minimum three thousand and that would be a great price."

François, in the meantime, was standing amid his art wearing his green ostrich leather vest, his pipe between his teeth, and a glass of white wine in his hand. He stood beside the only other artist selling art there, a beautiful woman named Marie-Annick who did lovely pastels of fish.

"Did you catch anything this morning, James?" François asked me.

"No, not a bite."

"I love fishing in the Seine but only in the summer," François shared. "I never catch fish with Pierre this time of year."

Fishermen walked by the art mounted on stands and hanging on the walls, but only François's friends stopped to look for very long. "They're not interested in art," said Marie-Annick. François poured her a glass of wine.

"It's true," François confirmed, opening another bottle. "They're only interested in the latest technology in rods and reels. They've lost their purpose; they've become too removed from real fishing. I love fly-fishing, but they should try bait once in a while. It's more tactile." He took out a pouch, stuffed his pipe, and looked vacantly at the room, pretending that he hadn't just made a philosophical statement or was even capable of making one.

I left François and Marie-Annick briefly to look at a sculpture he had done of a big *silure.* Staring into its mouth, which was wide enough to accommodate a dinner platter, I began to enact the ritual in my mind of catching one, as prehistoric men once made drawings of the hunt and their prey on the walls of caves the day before leaving to kill.

"The fishing should be good tomorrow," Pierre commented, stopping by to share a glass of wine. "It's raining right now. The river must be rising, and the strong current will force the bream to move into the eddy." He drank his wine. "The big *silure* will follow. When the bream are in thick, there are so many you can't bring a lure through the water without snagging one. The *silure* get in a frenzy. They herd the bream and sometimes push them to the surface, where they stun them with their giant tails and eat them." Pierre saw in my excited glance that his enthusiasm had been successfully contagious. "We'll fish every morning until we get one," he assured me.

The next morning I walked to Ile Saint-Louis alone to go fishing, and passed only two people on the way: green-uniformed workers picking up garbage in front of Notre-Dame. Pierre came on his bicycle an hour later. The bells on the cathedral echoed off the river.

I was leaning against a black lamppost when I got a tug on my lure, but all that I reeled in was my lure with silvery fish scales stuck to the hook. "The bream are in," I heard Pierre say to himself. "The *silure* should be too." But they did not come to my lure that morning.

"Tomorrow we will get serious," Pierre promised.

At the expo that day, the final day of the show, I was consumed by François's sculpture of the salmon, *le grand bécard vainqueur.* It assured its viewer that the fish was a worthy adversary.

After eating lunch, Pierre and I walked in the rain to a fish market on rue Mazarin to buy mackerel and squid for *silure* bait.

"When I decide to do something, I do it seriously," Pierre said.

That night we prepared all the gear—six rods, hooks and weights, folding chairs, an umbrella, and wading boots—and stowed it in Pierre's van.

"Don't you want to refrigerate that?" I asked Pierre as he loaded the mackerel and squid into a canvas bag.

"No," he replied, "I want it to stink."

By seven the next morning we were on Ile Saint-Louis. By seven-thirty all six lines were rigged and set in the water.

At that hour in February it was still dark; the lampposts were lit and the city was quiet. A light drizzle fell on my face as I stretched out on a marble bench and closed my eyes. After a few minutes I lifted my head to see if Pierre was still there. I looked around, and in my half-awake and half-asleep state, the current of the river going by the island gave the illusion that I was aboard a moving ship. I also had the sensation that the ship was sinking, because the river was almost visibly rising from seven days of on-and-off rain.

Near sunrise, Pierre picked up one of the rods and jigged for zander. On the second cast, he retrieved the lure with a fish scale—slightly oblong with a mother-of-pearl brilliance—stuck on the hook. "Bream!" he exclaimed prophetically. "It's a good sign for *silure.*"

At this pronouncement, I stared at the lines with new hope, which lasted for about five minutes. My eyes were averted to traffic on the voie Georges Pompidou—the morning migration of Parisians to the west side of the city.

When there was enough light to read by, Pierre pulled out a day-old copy of *Le Monde.* He read for two minutes, and then looked up at the lines. "Shit!" he yelled. "Where are the *silure?* We have done everything right here, and the conditions are perfect. If the rain keeps up they will have to close the expressway. Look, it'll be just a couple feet before it's underwater!" He walked over to our bait supply and, chopping several mackerel in pieces, started chumming the eddy. "Well, at least we are not stuck in traffic," he conceded, his bloodied hands throwing chunks of fish into the river.

"Shi-it, where are the *silure?*"

Several minutes passed.

Pierre laughed as he looked toward the right bank. "There is one now." He pointed to a long barge on the cabin of which was printed in block letters, *Le Silure.* It was one of two sanitation barges for the city of Paris that clean the dog feces off the quays and scoop up floating debris, named after the bottom-feeding omnivore we were in pursuit of.

As the morning grew brighter, joggers ran by us around the tip of the island. People walked their dogs on leashes. The dogs sniffed our bait and tangled their feet in our lines. Shortly, the sanitation barge moored to the island with a large metal ring. A man in a green suit jumped out and began to hose down the cobbles. The bells at Notre-Dame and other nearby churches sounded nine-thirty.

## MEANWHILE IN ROUEN—MONET AND THE FISH IN HIS LILY POND

When I next talked to Yannid in Rouen she had just finished stitching a man who was hemorrhaging badly after a car accident. The medical students practiced on cases that were more or less terminal. "He died underneath me," Yannid lamented. "That's the first time that happened."

I had returned to Rouen to await the spring and to be with Yannid.

Every day that went by I noticed the sun was stronger, and the tips of the willow branches were greening. The early leaves on the chestnut trees along the river spread from the branches in formations resembling the soft paws of a pouncing snow cat. I told myself that I would not let the emergence of spring pass without watching it closely.

I spent many mild mornings by the river reading or drawing in my sketchbook. Sometimes I was caught in a brief downpour, only to return to the apartment wet. Yannid would see my book warped from the rain and ask why I had not sought refuge under the bridge by the river. I never had a good answer.

Two weeks into April, Yannid kept her promise to show me the Norman countryside in spring. We packed weekend bags and drove to her aunt's house, a mansion near Etrépagny. Her aunt, who lived alone, was away and had left us the key to the house with instructions to feed her dog and water the garden.

When we arrived at the large redbrick home, Yannid took me inside and led me upstairs to a guest room on the third floor where she used to stay in summer as a girl. After a time lying in the bed and then napping with the window open, we decided to put aside serious talk about the future and take a walk.

Yannid led me to her aunt's garden, where she took off her shoes and walked barefoot on the soil between rows of gooseberries. I thought then, just by the way she stood there, that Yannid would not mind very much when I left France to continue on my trip.

That night for dinner we cooked a duck marinated in raspberry wine. Neither of us ate very much, though we managed to share two bottles of good Bordeaux from her aunt's cellar. After the wine, we strolled outside in the dark under a nearly full moon. Standing at the edge of the garden again, Yannid kissed me. We returned to the house holding hands and reaffirmed our affection for each other in the guest room on the third floor.

The next morning Yannid rose early and collected brown eggs from the chicken coop by the garden. She cooked a large omelette and we ate heartily without saying much.

After breakfast we drove to Giverny to see the house and garden of the painter Claude Monet. The day was warm and held fragrances in the air. Yannid stopped to smell a blossom on our walk through the garden to Monet's pond of water lilies.

The pond, we discovered, was fed by a channel from a cold spring-fed trout stream. Yannid sat by the stream to enjoy its refreshing coolness and I sat next to her. She leaned over and kissed me. We had talked little that day so it surprised me when she spoke.

"Hey, there's a fish down there," she said, glancing into the water.

"Where?" I asked, for I couldn't see it at first.

"Below us there."

"You're right," I said, and took Yannid's hand and kissed it. "It's a trout."

I nestled my face in the crook of her neck and thought about driving back to Rouen for my rod, but it was a nice relief just to watch the trout and not have to catch it.

"It's nice just to watch it," Yannid noted.

"I was thinking the same thing," I agreed. We walked back through the garden to see Monet's house.

It was clear from looking at his paintings, especially his late work, that Monet saw infinite possibilities in depicting the way that water abstracted the willows and the sky and wove all their colors into its reflections. The walls of his home were hung almost exclusively with Japanese prints (many by Hiroshige), almost all of which depicted water, fish, or fishermen.

"Why didn't he paint the fish?" I asked Yannid later that day as we stared at several large carp in the lily pond.

"Only you would think of that," Yannid said. "Maybe he had enough to keep him busy on the surface. He knew he had to decide what to paint, and then master what he chose."

On the way back to her aunt's home, Yannid and I stopped at a fish market in Etrepagny and bought some fresh cod and squid.

When we arrived at the house Yannid opened a bottle of muscadet that had been chilling in the fridge. There was a cool musty flavor to the tall kitchen, which made Yannid's skin feel cool on my hand. I kissed her hand and held it between mine.

I sautéed the squid in olive oil with lots of garlic and we baked the cod in the oven with fresh vegetables. Yannid brought the wine out on the porch, where we had set a small round table with forks and plates and our meal. I lit a candle.

"What are you going to do when you leave France?" Yannid asked. I could see that she was getting sad and I heard a touch of vibrato in her voice. "I'm trying to be rational about this." A tear rolled down her cheek. "I'm lonely, you know. You're leaving this summer, and I want you to be happy and free. But I don't want you to go."

I kissed her cold cheek, now wet with tears. A cool breeze blew from the direction of the garden. "Let's clean up and go to bed," she said.

## An Exhibition of Fish

One among Pierre's circle of fishing friends was an elderly gentleman named André Schoeller. André always wore a tie, even when he fished, and never passed up an opportunity to show his talents as a raconteur. He shared memories of being a boy in Paris during the Nazi occupation, hinted at his relationship as a young man with the singer Edith Piaf and his friendship with Picasso, spoke of the record pike he had taken in his pond in Normandy, and also of the health of the Seine.

An art dealer for the better part of his life, André was in the midst of organizing an exhibition of paintings and sculptures of fish and fishing scenes done by painters, living and dead, who fished. To Schoeller, the crowning jewel of artists who had rendered trout on canvas was the French realist Gustave Courbet. He wanted badly to borrow Courbet's painting *La Truite* from the Musée d'Orsay to be the cornerstone of the exhibition, but even with his connections this was difficult. In the event that he could not borrow the painting, he had arranged for the next best thing, to hold the exhibition nearby the museum, on 13, quai de Conti, at the Galerie Larock-Granoff.

Pierre Larock, André's personal friend who owned the gallery, was famous for having inherited the largest private collection of Monet paintings from his aunt Katya, Monet's niece. Like Schoeller, Larock was a fisherman, and no doubt the idea for the exhibition took form over a lunch at their private pike-fishing pond in Normandy. Both François and Marie-Annick, whom I had previously met, were to be included in the exhibition as living artists, and I was asked to participate as well; dead artists included were Rebeyrolle, Messagier, and Miró, among others. The exhibition was arranged and a date for the opening was set.

In the meantime I had a trip planned to eastern France to visit and fish for trout with a friend of Pierre's, Philippe Boisson. I had been told that the trout—locally known as the *truite zébrée*—and the streams there were exquisite, especially the river Loue near the Swiss border where Philippe lived.

# Philippe Boisson and Life in Chenecey-Buillon

Philippe picked me up at the train station in Besançon, and we drove together in his red diesel Citroën to his apartment in the village of Chenecey-Buillon. He lived there with his young wife-to-be, Katy, on the second floor of an old stone farmhouse within earshot of the currents of the river Loue, where he had grown up fly-fishing.

Katy was petite, good humored, and pretty, and Philippe was a handsome left-handed fly fisher with a bump on his nose; as a couple they radiated contentedness and good health. She was a medical student, like Yannid, but found time to fish with Philippe between rounds at the hospital in Besançon.

There was not much to see in Chenecey-Buillon if you weren't interested in fishing or natural beauty. Besides a stone bridge, a bakery, a small inn with a bar, and a church, the village was an open meadow with red poppies. To those with rapid temperaments, a life there could be considered dull, but I doubted that the word had ever entered Philippe's mind.

His friends hung out at the boardinghouse and bar by the river. The proprietor was tall, wore a full black mustache, and knew everyone's name, even mine, shortly. Before I had spoken to him, he greeted us at the table with two beers on his tray.

"Show him the trout," the proprietor called proudly. When we had finished our beers, Philippe walked me to the dining room in the inn, where the skin mount of a large brown trout was displayed over the mantel of a hearth. The trout was sixteen pounds, Philippe explained, pointing out its enormous head and formidable teeth. It was different from any other trout I'd seen, striped at intervals with dark vertical bands. "The locals," he explained, "call this *la truite zébrée.* This fish was hooked and landed by my best friend, Norbert—the third largest ever taken in France on the fly. He caught it on a size-sixteen pheasant-tail nymph in August when the river was low and clear, on line with a breaking strength of only two and a half pounds." Like few other places in the world—Livingston, Montana, or Stockbridge, England—a skilled fly fisherman in Chenecey-Buillon was respected as nothing short of a virtuoso.

The afternoon I arrived, Philippe gave me his lucky fishing hat to wear. We drove with Katy on a dirt road through a large farm along the river, until we came to an emerald pool, the surface dimpled with the rises of feeding trout. Katy was on the water immediately, downstream of us, making long graceful casts across the pool.

As I prepared to fish, Philippe kindly affirmed that the flies in my boxes were useless and handed me the precise fly for the occasion. I wondered some about his tackle too, especially the enormous net that hung over his shoulder, nearly two and a half feet wide. Then I remembered the big trout at the boardinghouse. I began to cast and he gave instructions in French.

"You must have no drag on the fly," Philippe advised, rubbing the bridge of his nose, "and then maybe, *maybe,* the trout will take."

I tried several casts to a rising fish at the tail of a pool, trying to let the fly drift as if it were a natural insect free of my line. On every cast my fly dragged before it reached the fish. "It's been a long winter," I yelled to Philippe over the sound of running water, and gave up, frustrated.

Philippe took my place in the river, made ten casts from the same spot, and hooked six trout. He landed only one because of his haste to show me what the trout in his river looked like. He held it out for me to see. It was true, the trout had the zebra stripes like he had said. "It's your turn, James," Philippe said, handing me his rod.

"Approach the fish slowly and get as close as you can before casting." As the sun was setting over the field of foot-high corn, I hooked my first *zébrée.*

I learned that the meaning of *fishing* for Philippe was waiting and observing.

Every morning of my stay with him and Katy, we sat on the terrace in front of the bar overlooking the river. The proprietor would bring us two *cafés.* Philippe unwrapped the paper from his cube of sugar, never taking his eyes off the surface of the river. He continued to watch the currents as he stirred the coffee with his spoon. The proprietor wiped the empty tables of morning dew with a blue towel.

"The fishing is very easy now," said Philippe, by which he meant as compared to the fishing in August, when the river was low and clear and the trout were fussy. The fishing was not easy for me, though. Philippe used very specialized flies he tied himself and fished with leaders of clear monofilament up to six or seven meters. He stalked the fish with such care that by the time he cast he was sometimes within a rod's length of them.

He was a magician at spotting trout over the light emerald gravel. If he stared long enough into the water it was inevitable that a fish would appear. Fishing with *mouche sèche,* dry fly, and on the surface was of no interest to Philippe, because he was interested in catching only the biggest trout.

"All the big fish, over three kilos, are caught on nymphs," he declared.

At times when we fished, we waited for trout so long that I thought I could see the sun tanning my arm as we sat in the grass. We mixed the waiting with eating, baguettes and local Franche Comté cheese, and drinking cold *cidre doux.* When the light was not right for spotting fish we even napped, which was nice because the breezes were always fragrant in the Jura and made for pleasant dreaming.

One afternoon, walking Philippe's beat on the Loue, we came across a large sick trout. It was finning in a still, quiet eddy where a healthy trout would never lie. It had white fungus growing over its eyes and was probably blind.

"Don't move, *truite vieille, truite malade,*" Philippe said, wading out to where it held over the gravel. "Old sick trout," he said, slipping the net underneath him. "He was a seven-pound fish when he was healthy."

When held in the light, there were vestiges of gold on its broad sides. "Even in this condition it is a good specimen of *zébrée.* You see the stripes? Its *tête énorme,* and the big *nageoires?* It is Courbet's trout."

Philippe took me to Ornans one day to visit the house and studio of the nineteenth-century painter Gustave Courbet.

Courbet was born in the town of Ornans on the Loue, some miles upstream from where Philippe lived. Courbet was a fisherman and through the course of his career painted several oils of the native trout from the river, one of which hangs in the Musée d'Orsay in Paris. When I first saw Courbet's *Truite* in the Orsay, I didn't understand Pierre's or André's rapturous descriptions of it—it didn't look like any trout I had ever seen. The fish's colors were washed out, almost silvery, with a faint yellow cast. It had small irregular spots like cracked peppercorns, a black ventral fin as large as a sail, and an enormous and almost grotesque head. Only now, after I had seen a big trout from la Loue, could I appreciate Courbet's painting; it was true.

The centuries-old stone homes in Ornans seemed to grow from the river, their foundations in the currents, their terraces spilling over the river and hung with *jardinière.* Swallows dipped about and the occasional falcon could be seen chasing stoneflies from their perches on area cliffs. One of these homes, the one with the faded block letters BRASSERIE on its side, had once been the studio of Courbet.

The artist was a quiet hero; you didn't see the crowds here at Ornans that you did at Monet's home in Giverny. We walked down a narrow cobble street, rue Maison Courbet, to Courbet's door.

The three-story house was spacious; we walked up and down the creaky steps looking at several paintings. I read some passages here and there about Courbet's life and learned that during the months prior to painting his famous *Trout,* Courbet had served a jail sentence and suffered severe hemorrhoids.

> In the summer of 1872, back in Ornans to recuperate and to work freely, Courbet painted a different kind of *real allegory* of his experience. Though he had obviously fished many times before, he had never used fish as a theme as he had other game. Now struck perhaps by the fish as creature who is *caught* and who struggles vainly against his captor, he paints them: first in a more traditional way, as dead game, and then even more strikingly as a kind of self-portrait, inscribed with the phrase *in vinculi fecit* (made in chains).[1]

Philippe and I drove upstream to a bridge over la Loue where Courbet often went to watch trout.

We searched the bushes near the bridge for big stoneflies. Holding them by their abdomens we flicked the insects' heads with our middle fingers and then tossed them, stunned, off the bridge into the feeding zones of the big trout below. The trout watched the crippled stoneflies as they hit the water and floated downstream. Usually they took them with big sucking swirls, but sometimes they just rose and touched their noses to them without eating. *"Il le refuse!"* Philippe yelled.

[1]Sarah Faunce, *Courbet.* New York: Abrams, 1993.

## La Bienne

Philippe showed me all of his favorite views of the nearby river valleys—the Doubs at Goumois, the monastery by the Dessoubre, the turn in the river by the big mansion on la Loue. But the Bienne, above all, seemed to be Philippe's most secret trout river in the Jura.

Part of the attraction of the Bienne for Philippe was its difficulty. He also had not been fishing it that long; therefore, like a new love, it held the allure of the unknown. By contrast, he knew just about every fish on the Loue near his home.

"There are fewer fish in the Bienne," Philippe remarked, comparing the two rivers, with a tone as if to say that was good "not because more people fish it; there just aren't that many. But the ones that are there are big and difficult to catch."

We parked on the first bridge over the Bienne and stepped out of the car to take a look. Upstream of the bridge, Philippe spotted a good trout.

"Do you see it?" Philippe asked me, "it's next to the willow with the twisted branch, on top of the white rock." Philippe made a mental note of the fish's position. He took his fly rod out of the red Citroën and strung the line through the guides.

With the rod in the crook of his arm, Philippe inspected his flies in the battered metal boxes he kept in his vest. He picked out a small pheasant-tail nymph with an orange head and tied it onto his line. The wind was up and agitated the river's surface.

Philippe walked to the river and positioned himself behind the fish. He stood there one hour, waiting for the sun to come out of the clouds so that he might be able to see the fish again. I waited on the bridge above him.

He had nothing to aid him in the wait, neither a cigarette to smoke nor a piece of grass to chew on. I wondered what he was thinking about. When the sun did come out, the big fish was no longer there.

By the time we arrived at the next bridge the sky had cleared. Under a tree that overhung the left bank of the river, far below, Philippe spotted an enormous trout, probably six pounds. Its head was facing downstream, into the circling current of a large eddy. "It's the largest trout I've seen this spring," he whispered, as if the fish could hear us. "If I catch it I'll buy a bottle of champagne and we'll drink it."

If Philippe could position himself to cast under the overhanging tree without spooking the trout, the rest would be easy; the trout would take the fly and he would tire it and kill it in the deep green water under the bridge. It took him forty minutes to clamber down the hill and approach the fish. When he was close enough to cast, though, he saw that fungus had attacked its head and sides. "*Truite malade,*" he muttered, and walked back to the bridge.

We ate lunch out of the back of the red Citroën. Philippe was impatient to get back to the river. It was well into the afternoon and still he had not caught a fish.

When we arrived at the next spot the wind was up and it was difficult to spot fish in the uneven water over the flat yellow gravel. We sat on a rock above some briars overlooking the river. The clouds passed over and Philippe stared into the yellow-green water, looking for any movement or incongruence beneath the current.

Philippe touched his neck, red with sun, and let his rod lay limp in the briars. He shifted, stretched his spine, even picked up a pebble and flicked it in the tall grass along the bank. Cars passed behind him on the road and long shadows crept across the stones where he sat. I was passing into sleep.

Before Philippe had seen the cruising fish he had sensed it coming and was up on his feet. The trout he saw was big, moving upstream toward him, and he had slid down through the briars to the bank to prepare himself for their meeting. He changed the fly at the end of his leader, and by the time he did so, the fish was within casting range.

The trajectory was perfect, the fish sipped in the fly, and with a swift strike it was hooked.

Philippe ran down the river after the fish, holding his rod high, his arm extended above his head. One hundred yards downstream the gravel flats ended in a deep hole. He had landed the fish in his mind prematurely, though. The line broke and he held his head in his hands pulling air through his clenched teeth. He made one guttural curse at the passing clouds and then climbed through the briars to the car.

## CRAYFISHING IN THE *BOIS*

Fishing in the Seine was not the only option for one who chose to subsist as a hunter-gatherer in Paris. In his Paris memoir, *A Moveable Feast,* Hemingway wrote of killing pigeons with his trusty slingshot and making meals of them. He would not have bothered, perhaps, had he known of the prolific crayfish population in the ponds of the Bois de Boulogne in western Paris.

On my return to Paris, Pierre invited me to join him and his family on one of their mid-spring rituals, crayfishing in the *bois.*

The *bois* was a verdant wood of willow, chestnut, and plane trees. By Pierre's accounts the *écrevisse* were plentiful, large, easy to catch, and delicious in a cream-and-butter sauce.

*Fishing the Seine, off Ile Saint-Louis, Paris.*

Inside the old *bordel,* Carole was preparing food in baskets for our outing. The three children, May Fario, Venice, and Marlin, were running up and down the stairs fighting and yelling at one another. Pierre was in his cave, hastily filling a bag with the crayfish nets.

"This is the only way I can survive living in downtown Paris," he noted, "to get out once in a while."

After fighting through traffic in Pierre's van, we arrived in the *bois,* where the heat was dampened by the shade trees and the crowds had vaporized. Parisians in swimsuits sunbathed on the grass, and in quiet corners they lay out nude.

The collapsible crayfish traps were made of nylon netting pulled over concentric wire circles. The traps flattened when set on the mud-bottom of the pond, but when you pulled them up they were basin shaped.

May Fario and Venice were in charge of two traps each. They lowered their traps baited with chicken wings to the bottom and tied them by a string to trees on the bank. The best spot to set the traps was by a little bridge at the north end of the pond. They caught many crayfish, olive colored on the backs with orange legs and underclaws, and put them in a creel of woven reeds. Pierre set traps too, but never checked them; he fell asleep on the grass with a straw hat covering his face.

Marlin was too young to set his own traps. Instead, he raced naked up and down the path along the pond with a *casquette* on his head, charging pigeons with a sword of green willow. Meanwhile, Carole lit a cigarette and prepared lunch, unwrapping a checkered cloth on a picnic table.

We ate sugar melon and ham with baguettes and drank ice-cold beer. After lunch, Pierre went back to sleep. I stood with May Fario on the bank. She was an attractive and astute fourteen, with dark brown hair and freckles.

"He always does that," she complained, looking to her sleeping father. "But he's been better since Marlin was born." Just then, May

Fario was pulling up one of her traps and I noticed, as it emerged from the dark water, that a condom had settled in the bottom of it. "I'll reset this one," I said to her.

Carole seemed more comfortable here in the *bois.* As she smoked she shared childhood stories about crayfishing with her father on the shore of Lake Michigan, where she grew up. Marlin whipped his sword at pigeons and then sat with his bare butt on the picnic bench to look at the captured crayfish in the creel.

"Be careful, Marlin," Carole called.

*"Eclwewisse, eclwewisse,"* he repeated, reaching with his fingers into the basket of olive-and-orange-colored crayfish. They snapped at his fingers. Venice came up to the creel to show Marlin how to hold the crayfish.

When we had caught three dozen large crayfish, enough for dinner, we wound the lines around the traps, gathered the food, and returned through the congested city to rue Mazet to unload the gear.

Carole boiled the crayfish in salted water until they turned a savory red color. Then the crayfish were steeped in a sauce of butter and cream.

"You don't know how to eat them?" Pierre asked me when we sat down to eat and he saw my puzzled look. "Don't you have them in Connecticut? You take them like this with your hand and twist them like this and suck the meat out," he said, smiling, and looked at Marlin. "You must make as much noise as possible, *slurp, slurp,* but the best is this with the bread." He soaked up the creamy pink sauce with baguette hunks and poured more wine in my glass. "It's a good chardonnay," he said.

The children looked at Pierre as if he were crazy and laughed. Pierre and Carole argued over what wine Hemingway drank with *goujon,* a small sweet fish often fried whole, whether Sancerre or Chablis. Pierre corrected himself, "No, I'm sure it was muscadet."

# Fishing the Most Beautiful Pool in the World

The home of the artist–tax inspector François Calmejane was 41, rue de Seine, not far from Pierre's office.

"You will love François's apartment," Pierre told me as we waited outside his door. "We will have dinner there tonight after the fishing."

The street was silent except for the sound of clinking glasses, clanking silverware, and muffled conversation from La Palette restaurant. It was Saturday evening.

This part of the sixth arrondissement was known for harboring artistic and creative, sometimes eccentric, people. (Picasso painted *Guernica* in his studio here.) Even so, it was not every day that you saw a man step from a door in the stone façade dressed in fishing clothes.

François held his pipe handsomely in his teeth. A knowing smile crept across his face. He knew he looked ridiculous. He was wearing a bowler hat, a flannel shirt and jeans, and his fishing tackle was strapped to his body by old leather belts tied together so comfortably that it seemed he was never without them—a net, a box with bait, a creel, a tackle bag, and rods, each rigged with hooks and heavy lead sinkers.

"'Jour, Pierre," François said, and they shook hands. I took François's creel and Pierre took his net. In front of La Palette, where people were taking their dinners, François bent down to pick up a copy of the *Financial Times* from the gutter, its pink paper smeared with melted ice cream. He handed it to Pierre and Pierre handed it to me.

"To hold the eels, James," Pierre explained.

François led us through the people dining at the outdoor tables of La Palette. *"Bonjour, pêcheur,"* called the waiter, as if he'd seen him pass many times. The people dining laughed at him. Smoke curled from under the fisherman's hat, mingling with theirs.

The fisherman headed for Pont Neuf and the park at the tip of the island. He walked by the gilt statue of Henri IV and down the dark stairs under Pont Neuf. An Englishman brushed the *pêcheur* in the narrow passage and laughed to his companion. "You'd think the bloke was fishing for a year with all that gear."

When we reached the Seine, we saw our fishing spot at the tip of the island covered with people. They had come to enjoy the twilight by the water, gathered in groups all over the island, picnicking with dinner and wine under the thick, arching willow tree, laughing, playing drums and guitars among the roses. It was a loud, happy circus, in the midst of which the *pêcheur* assembled his fishing rods.

He laid out his tackle indelicately, dropping his bait box on a German man's leg, nearly spilling all the worms on the German lady beside him. The couple took their wine and moved farther from the fisherman. François stuffed his pipe and acknowledged his territorial gain with a shrug. Real estate was dear on the island on a Saturday evening in late May, and there is little wonder—it is one of the lovelier places in Paris to watch the sunset. Pierre put the sentiment in fisherman's terms.

"It's the most beautiful pool in the world," he said.

François baited one of his lines with a big nightcrawler. Swinging the long fiberglass rod, five-ounce sinker, and worm neatly between the heads of couples engaged at the mouth, he flung his bait into the middle of the river. It landed with a solid sound, *ploof.* He handed me the rod.

I put the rod down on the cobbles and the current drew the line taut. François looked up at me under his hat and took his pipe in hand.

"We lost a rod last year that way," Pierre remembered. "A big bream pulled it into the water but we were able to recover the line by casting a lure over it and retrieving it by hand. We got the rod back and the fish."

Along the Left Bank were moored fishing boats in motley colors and some clunky sailboats with wooden masts. Passing by at intervals were the tourist boats, the *bâteaux mouches,* some open to the air and some enclosed. Some carried couples dining over tablecloths. Their wakes pulled at our lines and bent our rods. The history of the river and the architecture of Paris were announced by intercom in four languages. The people on the boats waved to the people on the island.

"*Bonne pêche!*" one man yelled. François scowled.

"In France, *bonne pêche* or *bonne chasse* is considered a curse," Pierre explained, taking a chunk of cheese from François's bag and procuring a knife to cut it. "I've seen it happen dozens of times; you wish someone good fishing and they catch nothing—nothing! If you're going to say anything, it is better to say *bonne chance* or *bon voyage.*"

That was not the case this evening, though, because shortly thereafter one of the rods bent with a fish. The people sitting on their blankets couldn't believe the crazy *pêcheur* had actually hooked one.

"He's hooked the bottom," said the German to his lady. Moments later, François hauled a two-foot eel flipping up the inclined wall and onto the lap of another young lady, who was also German, a beautiful blond wearing a white dress. She spilled her glass of white wine and gave the *pêcheur* a nasty look with icy blue eyes. But François passed the cries of *Scheisse* to the big overhanging willow, which passed them to the breeze that blew from the sea at Le Havre.

A sedentary poet with an open book of Baudelaire, stationary as a bow ornament on a ship at the very tip of the island, uncrossed his legs and stood up wide-eyed as if he'd never seen a living creature pulled from the water.

*"Incroyable,"* the poet cried out, *"incroyable!"*

I had been flipping through the sticky *Financial Times* and admiring the way the descending sun played rose on the pink paper, but had to give up the "companies and markets" because the *pêcheur* demanded it. He took it with his big hands and laid it on the eel, and as the eel slipped through, it covered the stock-market tables in a thick white slime. He grabbed the eel's head and it wound its tail around the sleeve of his flannel coat—already encrusted with the dry crystallized mucus of a thousand eels.

The drummers looked on but continued their play confidently, as if they'd conjured the snake-fish from the Seine. François slid the eel into the wicker creel and drew the brass latch sound. He admitted to me as he rebaited his line and settled in his seat that he too had been surprised when, fifteen years before, he pulled his first eel from the Seine.

François then took a bottle of red wine from his canvas satchel and handed it to me to pull the cork. He bottled it himself and the label was his own design, a simple abstract sketch in ink. *"Moins cher,"* he said.

The sun was just below the metal grid of Pont des Arts and silhouetted the people viewing the display there of massive papier-mâché, burlap, and fiberglass sculptures of men and horses that a Senegalese man had created to represent the battle of the Little Bighorn.

General Custer stood tall with gun in hand, careful not to fall into the Seine. He took his last stand west of us, between felled horses tossing their massive hooves and tails.

Now the passing boats with their shining lights reflected in Pierre's glasses and shone red through the numerous wine bottles surrounded by their circles of drinkers. Some people had brought candles that they lit and stuck in the mouths of empty bottles or adhered with hot wax to the cobbles. François showed a nervous alertness and put down his glass. *"J'ai une touche,"* he said, for he did have a bite. Another eel came winding serpentine up the wall and into François's creel. He lit his pipe with a fish-shaped lighter he had carved that hung by a string from his neck. Bells rang from Notre-Dame cathedral. The Louvre was red just before the sun went down.

François shared his fifth-floor apartment at 41, rue de Seine with his wife, his daughters Euridice and Penelope, and his obsession with fish. The fishing gear was stowed and the creel full of eels set down.

On the way back from the river, Pierre had gone to his own apartment on rue Dauphine to get a loaf of bread and a bottle of champagne. By the time he returned I had glimpsed the crowded room where François introduced guests to the workings of his mind.

François made things from found objects, he sculpted, he sawed, he welded, he painted. His studio was an attic space up a small stair, too small for a grown man to stand in. In the living room there were skin mounts and dried pieces of fish he had caught in various parts of the world—a dorado from the Caribbean, a taimen from Mongolia—and on the brick-and-stone floor were plaster casts of the largest fish he had caught in the Seine, a five-kilo bream, a seven-kilo zander. There were large nightmarish fishing flies he had welded from metal fragments—the tines of a pitchfork for legs, an oil-can abdomen, the spool of a fishing reel for a head. In the middle of the room was a dark wooden table in the shape of a fish, which François had carved from slabs of apple wood attached with metal strips and screws.

On a shelf along the inside wall were clay and lead sculptures he had made, many of them representing his friends. Among them was a bronze of Pierre in the nude holding a big salmon by the tail, his body arched back, midsection thrust forward, a fishing rod for a penis. Hanging on the wall was the *pêcheur*'s self-portrait with a salmon from the Moy River in Ireland. Fish descended from the ceilings and swam up from the floors, leaped in stained glass from the windows.

"François is remarkable. What a fantastic place," cried Pierre, coming in through the door, his hand on the neck of the champagne bottle, the baguette wrapped in a paper bag under his arm.

The *pêcheur* had sautéed eel in olive oil and laid out some blood sausage grilled with potatoes. We ate off wooden plates on the table shaped like a fish and drank alternately champagne and *cahors.*

"I love potatoes," said François.

Some time in the early morning I stumbled back to Pierre's office to sleep and Pierre walked back to his home. Only moments later, it seemed, Pierre rang the office telephone to wake me. I had a headache and my tongue felt like a piece of salted cod.

Rue Dauphine was quiet and damp from rain. I walked over Pont Neuf down the stairs to the place du Vert Galant and to the tip of the island, where the *pêcheur* already had his lines in the water and stood puffing his pipe under the overhanging willow tree. "'Jour," François said, his breath visible in the cool morning. Pierre came to the tip of the island shortly after, his hands in his pockets, a hat on his head. I wondered if he had gone to sleep at all.

All the colorful boats were moored, collecting dew. When the sun started to creep up over Notre-Dame, men appeared to sop up the dew with rags and dry the seats for passengers. François had a golden bell on the tip of one of his rods and the morning light sparkled on it as it did on the gold of the dome of the Institut de France, shining against the sky, which grew gray with clouds. Custer and the Sioux made of papier-mâché were still fighting on Pont des Arts, now in a creeping mist. Then a light drizzle began to wet the cobbles on the island.

*"Regarde,"* François called when he saw a fish jump, *"belle gobage!"* Then he raised his rod and reeled in a good bream. The broad

fish bled as he took out the hook and slid it back down the wall into the water.

The first barge up the river was heaped with asphalt and hung low in the water. The first two men on the island were the cleanup crew who came to pick up all the empty wine bottles and scrape the wax from the cobbles with razors. The pigeons scavenged food between the melted candles. An old man walked among the roses with his dog.

The *pêcheur* brought in his lines. The drizzle became rain and the rain came harder and the gray zinc roof of the Hôtel de Ville was black.

## Sea Monsters

Spring was easing into summer and we'd had two solid weeks of rain, which brought the level of the river up to February levels and beyond. Pierre began to get excited, not about the opening of fish art that André had organized, but the fishing.

"With the flood we may have another chance for a *silure.* They will begin to chase the bream that congregate in the eddies behind the islands."

The date of the gallery opening had arrived. Marie-Annick contributed several watercolors and pastels of salmon, bream, and carp, and François had carved several trout and a pike from a dead apple tree. Pierre and I were out on the river fishing as they were setting up at the Larock-Granoff Gallery for the *vernissage.*

It was raining, but the air was warm and fragrant with spring, so it was pleasant to stand on Ile Saint-Louis with a fishing rod in hand. The river had risen swiftly and was now over the lip of the wall on the island. We waded in ankle-deep water as we cast our jigs into the eddy.

"It's strange," Pierre said, "but the river now just reminded me of a salmon river—like the Tweed should be at this time. The first fish should be starting to run." He wiped his gnomish nose with the back of his hand and reeled in his line. "I wish you had met Charles Ritz before he died. He was an elegant man, like André.

"Ritz spotted my talent as a fly caster and introduced me to salmon fishing when I was about your age. He took me to fish the remarkable Aaro of Norway," Pierre said, speaking in a soft tone that was uncharacteristic of him. "The Aaro is a very short river, less than one kilometer, that empties into Sognefjord. Ritz was very good friends then with the grandson of Count Tolstoy, Sasha, who had a lease on the water." Pierre laughed. "We had good times. I remember him telling Sasha one night about the Allier in France, the second longest Atlantic salmon river in the world, how beautiful a fishery it once had been. But even in Ritz's time the salmon were in trouble.

"I should take you to see the Allier. It is maybe my favorite river. In Saint Valiene de Chaz, on the bank of the river, there's a thirteenth-century stone church. The villagers built it to commemorate an early run of salmon that had saved them from starvation after a hard winter. In a normal year the fish reached the Saint Valiene by May—that year they were in by mid-March. The villagers thought it was a miracle! Maybe it was." Pierre looked into the Seine. "I think we may even catch a salmon there.

"Well," Pierre said, interrupting himself to look at his watch, "it's eight, I have to take Marlin to school. Keep fishing; I'll come back on my bicycle in an hour or so."

I had fished so many times for the notorious *silure* without catching one that I believed I never would. When Pierre left I lost all confidence, and put down my fishing rod and stretched out on the marble bench behind me. I was very tired, but I did not want Pierre to return and find me there asleep. "What, you have given up already?" I could hear him say. I reluctantly rose again, grabbed the rod, and cast a jig out into the eddy. I let the lure sink to the bottom and reeled it back slowly.

My line stopped; I had hooked something and it was moving, throbbing at the end of the line, a fish. At first it didn't feel that big, but then it came to the surface and boiled. It burrowed down like a mole in soft soil and then surfaced again, rolling, and pushing water like a drowning cow.

It moved in slow figure-eight patterns back and forth across the eddy. Moments later it surfaced again, slapping its tail and showing a bulbous white belly to the sky. I heard cries of amazement behind me and saw that I had attracted a small audience of people. One of them had his camera poised. A group of schoolgirls who spotted me fighting the fish as they crossed the Pont St.-Louis piled onto the tip of the island to watch. *"C'est gros!"* they exclaimed.

I fought the fish, careful not to put too much pressure on the line.

"Thirty kilos, *peut-être,*" I called to the collected crowd.

I tried to lift the fish from the bottom so that they could see it. By a clock that I could see through the branches of the twisted chestnut tree nearby, I saw I had been fighting the fish for ten minutes. Impatience was growing among the people on the tip of the île but soon I had the fish's mouth up to the lip of the wall. I was close enough to try to get my hand under his gill and land him—I reached out and touched the soft white skin under his jaw, trying to get my hand in its mouth. His tiny white eyes with dark pupils looked blankly at me as he lingered there—then he slapped his giant tail and slipped off the cement lip back into the river. The line made a singing sound as it spun off the reel.

A short black-haired man sat on the cement bench, hurriedly taking off his shoes, rolling up his pants so he could come to my aid. Showing great heroism, he waded into the ankle-deep water and stood beside me, staring into the opaque yellow depths off the lip of the island. When I reeled the fish's head up to the lip of the cement wall again, I handed him the rod. He walked slowly backward with it and I reached out over the edge of the island and put my hand in the fish's giant mouth. Grabbing it by its lower jaw, I dragged it up into the shallow water over the cobbles.

"*Dieu,*" said the man, his eyes wide.

Shortly, Pierre arrived at the tip of the île on his bicycle and pushed through the crowd. "You got your *silure,*" he said, seeing the fish, bending over to look into its small bead-shaped eyes. "I am happy. It is fifty pounds easy."

"It's the biggest fish I've ever caught," I said, smiling and looking at the big fish with admiration.

"And would you believe, they get much bigger," Pierre stated.

I put my hand under its gill and held it up high as if it were the severed head of Medusa, to show everyone who had stopped to see what lived in the river they strolled along every day. I lowered the *silure* to the water again and strung a long rope through its gill and out the mouth so I could secure it to the lamppost. It was alive in the water until we decided its fate.

Pierre suggested I keep the fish. "François will want to see it," he said, dialing him at the gallery with his cell phone. "He doesn't believe me that they exist. He's been wanting to catch one and eat it for years."

François was setting up his sculptures when we called, but dropped everything to come down to the river to see the *silure.* André came too, and when they arrived I pulled the *silure* from the water and the two of them kneeled beside it as if it were an altar. There was something classical and peculiar in seeing André dressed in his jacket and tie beside the large fish.

"*Quelle vache,*" what a cow, he repeated over and over like a gentle mantra, staring into its blank, black-eyed-pea eyes.

I asked Pierre to pose with the fish and I sketched him with the sea monster. "Yannid will love you when she sees what you have caught," Pierre said.

It was decided that the fish would be killed and eaten. Pierre took out his pocketknife and, laying the *silure* on a small spot of grass above the waterline, he stabbed it through its skull to kill it, and then cut open its belly, allowing masses of multicolored entrails, rounded and fluid like wind-filled sails, to spill onto the grass. Indeed, as André had said, it was a cow, as with the entrails came handfuls of tiny eggs attached by a mucous skein. François unfolded a large burlap sack and, stuffing the giant fish in, slung it over his shoulder and walked away.

The reception that night at the Larock-Granoff Gallery was standard as far as openings go, only there were not enough paper cups and bottles of champagne to go around. Yannid could not make it to the opening, news which would not have bothered me had it not been compounded with the remorse I felt in seeing the big fish killed.

I did not see the *silure* again until we went to François's apartment for drinks and dinner after the opening. He had hung it by the head in the doorway so that everyone entering could see it and, thusly, had made me into a kind of unexpected hero. Everyone was awestruck by the strange, almost mythical creature that had come out of the river in downtown Paris. François's young daughters, Euridice and Penelope, asked me how I came to catch it. I told them it struck my lure when I least expected it to, that I had caught it on the first cast after waking up from a nap on a bench.

"Maybe you never woke up," Penelope said.

"When I was a girl," began Euridice, smiling, "I told my father that I would marry the man who caught a *silure* of two meters and one-half." The one I had caught was about one meter and one-half, more than four feet long. "I don't think yours is big enough."

The wine flowed freely in the dark apartment, drunk with sausages and cheeses, and a sturgeon pâté François had made.

The next morning Pierre got a call from his friend Jean-François, who was flying from Paris to Dublin and had seen my picture in the Parisian daily paper. "There's a picture of your American friend holding a *silure* on the cover of *Le Parisien,*" Jean-François said. One of the witnesses of the catch that morning had been a journalist. The same picture of the *pêche au gros!*, as it was described, appeared days later in the national paper, *Le Figaro*, and subsequently in the *Nice Matin*. The caption read:

> *Insolite: Pêche au gros!*
>
> You don't need to go to the Caribbean Sea to rub shoulders with marine monsters. Here at 10:30 A.M. a fisherman pulled from the Seine a *silure,* a giant version of a catfish, estimated to be 25 kilos. A prize, fulfilled after a long fight with feet in the water, it was led to the tip of Ile Saint-Louis by . . . an American tourist.

## THE HAUTE SEINE—THE VINE LIFE

François, Pierre, and another friend of theirs whom they called Lulu planned to continue the festivities of the *vernissage* the next day by going fishing on the upper Seine for trout and grayling.

The Seine near its source, a two-hour drive from downtown Paris, was a beautiful spring-fed trout stream, winding peacefully through bucolic pastureland on the border of Champagne and Burgundy.

Driving us all in his van, François pulled over at the first vineyard we came to in Champagne, Veuve Cheurlin. It was not a large vineyard and they had no formal tasting room, so the four of us walked into the small office and asked the woman in attendance if we could taste some champagne.

First she brought out their reserve brut and poured us each a small glass.

"Yes, that is just okay," said Pierre, "can we try another?"

The woman brought out another bottle, peeled off the foil, and pulled the cork. "Yes, this is better," said Pierre, tasting it, "but let me try a little more of it." She poured him a small bit and he gulped it down.

"Is your reserve the best?" François said.

"No," she said, "our *cuvée prestige brut* is the most expensive." Pierre stared at her expectantly and she sent a boy to get a bottle. She opened it and poured a bit in each of our glasses. Pierre took out a pen and paper and did some figuring, wiping his mouth with his hand. "I'll take eighteen bottles of the prestige," he said, and the woman hid her smile.

With the van full of champagne we came into Burgundy. François turned off the road onto a dirt path through hay fields dotted with red poppies until we came to a small silver trailer in a field. He kept it there as a fishing shack by permission of the farmer; the trout stream was a short walk across a wide field.

Before lunch we walked to the Seine, here a fledgling river near its source. "The water is a little high," Pierre remarked, "the mayfly hatch will be late this season."

It had been a wet and cold spring but this felt like the first day of summer. After a bit of walking along the river and not seeing any fish, we all returned to the trailer to take off our boots and have some food and drink.

The small trailer in the farmer's field was a kind of mobile home with a stove and cabinets and a table with chairs to sit around. Its kitchen was stocked with wine, Pernod, plates, pots, and utensils. We had brought along a half dozen baguettes and lots of Roquefort,

Camembert, butter, melon, grapes, and ham. François had brought a piece of my *silure* and began to cook it in a pot on a small range with butter and shallots. Pierre opened a bottle of wine.

"I don't like Camembert," Pierre said, "but this Roquefort is fantastic! And it is really good with this *cahors.* It is a very strong red wine, perfect for Roquefort, you know."

Lulu carried on about the virtues of Camembert.

"Do you talk so much about cheese in America?" François asked.

"No," I joked, and poured myself more wine, "we just eat it."

"You know," Pierre said, "one of your American authors, Jim Harrison, I met him once in Paris—he talks and writes very well about French cheese."

Between us we drank five bottles of wine and then went fishing out in the sun. We managed to remember our fishing rods but forgot our purpose by the time we crossed the large field between the trailer and the river. Pierre lay down in the thigh-high grass. Lulu didn't make it to the river either, but dropped like a felled soldier. François made it to the bank and quietly lit his pipe under the shade of a chestnut tree. I took some casts in the river and then found my own place to rest, in the shade of a tall oak.

As I slept, the sun climbed higher in the sky, stealing my shade, and I woke lying in direct sun, my face flush and my body covered in sweat. I found a deep pool on the river, took off my clothes, and jumped in. Once I had cooled off I fell asleep in the shade again.

When I awoke I looked over at Pierre, still sleeping in the grass. It made me think of the painter Gustave Courbet and his two friends who used to make outings to fish and paint by the river Loue. There was a painting in the show at the Larock-Granoff Gallery by Courbet's friend Célice, depicting Courbet fishing with a cane pole off a steep bank. I didn't really like the painting, but it reminded me that it was nice to be on the water with friends.

"Il le refuse!" La truite zébrée, *Rivière de La Loue, eastern France.*

# Departing Paris

Before I left Paris on a train to southern Austria, there was one last thing I needed to do. I walked to 41, rue de Seine to visit François Calmejane.

He greeted me at the door wearing an apron, a pipe in his tobacco-stained teeth. I could not tell if he had been cooking or welding, though it did not matter, for whatever he'd been doing I knew he was creating. He was so happy when I told him I wanted to buy his sculpture of the salmon that I had first seen at the fly-fishing *salon, le grand bécard vainqueur,* that I thought the old tax inspector was going to cry. He took me in his arms, crying, "Oh, thank you, James, thank you," and said it pleased him not only for the money, but because he knew the *bécard* would be with a young man who was a fellow artist and that his work would be seen in America.

The sculpture was large, even with the stand removed, and I hoped François would find a secure way of shipping it to my father's home, where it would await my return. "I don't have a choice whether to buy it or not," I said to François. I wished more things were so clear in life as a trout stream or good art.

## Part II

# Johannes Schöffmann and the Trout of the Tigris River

Some years before, I had written to the eminent fish biologist Dr. Robert Behnke at Colorado State University in Fort Collins, asking if there were trout in the headwaters of the Tigris and Euphrates of eastern Turkey.

"There's only one man I know who has fished for trout in eastern Turkey," Behnke wrote me. "His name is Johannes Schöffmann and he lives in southern Austria. The first letter I received from him was in regards to a very rare and little-known trout of south central Turkey called *Platysalmo platycephalus.* The only other thing I know about him is that he's an accomplished and inquisitive scientist who has traveled widely in search of trout. As far as I know, he is not affiliated with any university. He is, in the true sense of the word, an amateur."

Johannes published his ichthyological findings in the Austrian scientific journal *Österreichs Fischerei.* He not only did his own original research, but helped others perform theirs. He kept in close contact with several scientists at universities in Europe, Asia, and America who depended on him to collect trout from remote sites for their research (not many people had fished for trout in the mountain climes of Turkey, Morocco, Iran, or Algeria). These biologists, in particular Louis Bernatchez of Laval University in Quebec, were working to create an evolutionary map of *Salmo trutta,* the brown trout, by studying its mitochondrial DNA. It was that species—by some sources the most genetically diverse vertebrate on the planet—that was the object of Johannes's *loucura.*

I was on a train that had just stopped in Salzburg, already thirteen hours from the Gare de l'Est in Paris and four hours from my destination, Sankt Veit an der Glan, Austria. I knew very little about what lay east of Paris, so I was excited to discover what I might see.

Through Behnke's introduction, I had begun a correspondence with Herr Schöffmann before I left home and I reread some of his letters on the train.

My questions to him were about the trout of Turkey, particularly in the headwaters of the Tigris and Euphrates, what they looked like, the nature of their habitat, and, on a different note, my concerns about the safety of traveling in eastern Turkey. I had not noticed until then, the cool night sliding by outside the train, that Johannes had never addressed my questions on safety, but stated only that he was planning a trip to southeast Turkey in search of trout in the headwaters of the Tigris River. He added that I was free to join him if I wished.

I did not know Herr Schöffmann's occupation, but somehow imagined him as a corporate man who wore a tie to work. I discovered something quite different. I was greeted at the train station by a man in khakis, dressed like a British explorer—I was the only passenger to disembark in his small town, Sankt Veit. It was not at first clear how we would communicate, as his English was poor and I spoke no German. After several false starts we settled on our best mutual language, which happened to be Spanish. Conversation came easier, though by no means fluidly.

Johannes lived near the station on the second floor of a beige plaster-faced apartment building. To get to the stairwell we had to pass through a room hazy with confection smells. The ground floor was a bakery, and Herr Schöffmann, it turns out, was a baker. The door on the second floor led into the kitchen, where we took off our shoes and where I first met Ida, Johannes's wife. He explained to her that I spoke no German, but Spanish. Ida laughed, her stout and plump body shaking, and explained to me that they had learned Spanish in Colombia, where Johannes had apprenticed for two years as a young pastry chef in one of the best restaurants in Bogotá.

Johannes was no longer young, was mostly bald, with a reddish aspect and a dusky mustache. We sat down at a table in their living room and Ida brought us each a bottle of Gösser beer with a glass. I *had* correctly imagined Johannes as a man of few words. We poured our beers and toasted, though at that point we knew not what we were toasting to.

"*Prost,*" he said, lifting his glass.

"*Prost,*" I repeated.

"I must sleep for a bit," Johannes said then, excusing himself. "I woke early; a baker's work life is nocturnal."

Ida sat down with me while I finished my beer. She lit a cigarette. Her complexion was dark and I thought she was Turkish maybe, though she explained her father was from India. "You look tired," she said, and showed me to a small room where I could rest. It was a girl's room, hung with posters of American teen idols. I must be displacing their daughter, I thought.

I slept too and did not see Johannes again until afternoon, when he woke me to share his plan.

We sat at the table in his living room and Johannes showed me several albums filled with photos from his travels. What struck me immediately, whether the photos were from Lebanon, Bulgaria, or Croatia, was that Johannes had been able to find local people to help his explorations wherever he went. His impromptu guides appeared beside him, an Arab, a Basque, a Kurd, a Berber, showing me that fishing was a powerful tool for reaching across languages and cultures.

As I flipped through the photos I asked Johannes questions. I soon realized that if I was looking for answers this exercise was futile.

"How do you catch the fish?" I asked Johannes.

"With my hands," he replied, which I thought was surely a joke.

"How did you find trout in these places, how do you communicate with the people? Did you have translators?"

"Why, you don't speak Turkish?" he questioned in his deep voice and laughed.

Without showing me a map he told me the object of our expedition, to catch a trout in the headwaters of the Tigris River. We were to drive several thousand miles from Austria almost to the Iraqi border in southeast Turkey to reach high tributaries flowing from the mountains. The Tigris trout were important to Johannes, foremost because he had never caught or seen them. There was only one published reference to their existence (in an early-twentieth-century paper by the Italian ichthyologist Enrico Tortonese), and no pictures or detailed descriptions of live specimens.

"I think they will be different from other brown trout," he predicted.

To understand Johannes's theory on why the Tigris trout would be special, I had to understand a bit of the theory of how trout evolved. Ancestral trout thousands of years ago invaded rivers from the ocean during the end of glacial periods when there were significant inland waterways from melting ice that no longer exist. When the meltwater retreated to the ocean, the ancestral trout that had invaded from the ocean were stranded in different streams and evolved into the different types of trout we have today. In eastern Turkey, trout invaded rivers from the Black Sea, the Caspian, and the Mediterranean, the headwaters of which, in certain places near mountain divides, came very close to touching one another and also to connecting to the headwater tributaries of the Tigris River, which flow all the way to the Indian Ocean. It is thought that the Indian Ocean was too warm to sustain the ancestral trout, therefore the trout in the headwaters of the Tigris must have come from somewhere else; they swam over the mountain passes from the Caspian and Mediterranean, and perhaps even Black Sea basins, when there was enough water from glacial melt to do so, and have since been stranded there for thousands of years.

Searching for this mysterious trout would be an adventure. The journey would be long and dangerous. The streams where the trout lived were in the most politically turbulent part of the country—near the Iraqi border where Turkish military were actively fighting the Kurdish separatists. All this seemed to please Johannes immensely. It all made me a little nervous.

Johannes's town, Sankt Veit an der Glan, was said to have ten thousand people and eighty pubs. That seems an astonishing ratio, but I soon learned how they all stayed in business. Over the course of my first days in Austria, I think that Johannes and I visited most of them and I soon developed a capacity for great amounts of beer. Sometimes we were joined by Ida or Johannes's two children, both younger than myself, Mariela and Benedikt, but usually it was just Johannes and me. It was through his friends, not himself, that I learned about Johannes.

"Johannes is crazy," one of his friends told me as we drank our Villacher beer, "he risks his life to find trout. Don't you notice how he wears those fatiguelike khakis? He's a real explorer."

Johannes, too, painted, as we discovered over drinks. He was also, like me, a Gemini, left-handed, and shared a curiosity for all natural things. His friends confirmed, in a combination of English and German, what I had already realized—that Johannes was not just a simple baker. He had traveled widely and learned pieces of far-off languages; he was the prodigal son of Sankt Veit.

Our first outing to a river together was on a sunny Sunday morning four days after I'd arrived. Johannes, Ida, and I drove in his old Land Rover up into the mountains and over a long winding pass to Slovenia. The valley we came into was a soft velvety green with a milky emerald river flowing through, the Soča.

When we came to the floor of the valley, Johannes pulled onto a dirt road and parked at the edge of a thick wood by the river. On a tree in front of us was a sign—a fish with a black X over it.

I stepped out of the car to grab my fly rod in the trunk, but Johannes shook his head.

"Not here," he said. He pulled out, instead, a wet suit, and, stripping to his underwear in seconds, zipped himself inside the black neoprene. It covered every part of his body but his face. Ida stood beside him clutching a blue towel. The towel bulged with something hidden beneath it. They made for the river, Johannes walking with several kilos of lead on his waist belt, his mask, snorkel, and fins in his hand.

As Johannes submerged himself in the crystal-clear and frigid currents of the river, Ida unfolded the towel, revealing a small net. She handed it to him and he dove, though I could still see his distorted black form. In seconds he had returned to the bank with a small trout, smiling like a boy who had ensnared a butterfly. So, it was as he had told me; he did catch trout with his hands.

"This is a marble trout," Johannes said, his elbows propped up on the bank, the fish in his gloved hands. "It is native only to certain rivers draining the Adriatic Sea, like the Soča here." He let the little fish patterned with vermiculated sides slip back into the water.

Later that day, the three of us shared a bottle of chilled red wine and a plate of prosciutto and cheese at a local restaurant called the Gostilna Žvikar. It was a favorite stop for Johannes and Ida. We sat under umbrellas outside as a rainstorm swept through the valley. Purple thunderheads rumbled.

"My head hurts from diving in the cold water," Johannes said and laughed. "But diving is the only safe way to fish if a warden is around."

As we drank the wine we discussed the itinerary for our trip, which would take us from Austria to Italy the next day and then by ferry to Greece. Johannes revealed that oftentimes we would be searching for trout in national parks where fishing was closed, or without licenses.

"We will have to be *Schwarzfischers,*" Johannes said. Ida laughed.

"What is that?"

"*Schwarzfischer* means literally black fisher, it is German for poacher."

So we founded an international society to describe our cause, La Sociedad Internacional de Schwarzfischers. We would fish by any means possible and catch trout for art and science regardless of the regulations.

The next day our trip began, overland from Sankt Veit to the city of Trieste in northern Italy. From there we took a ferry twenty-five hours down the Adriatic Sea to Igoumenitsa, Greece. It was early the following morning that we came to the first river we would fish on our four-week trip, the Voïdomátis.

I sat in the backseat with notebooks, the tent, the vials of alcohol for specimen preservation, sleeping bags, and diving equipment. In Trieste (the birthplace of my mother's father) I had bought my own wet suit and diving equipment.

Green fields of corn and wheat and acres of sunflowers spread to the foothills of the Voreia mountains that skirted the Greek-Albanian border.

Johannes lit a cigarette.

"In Tiranë, Albania, my friend Peter and I rented a car, it had no brakes," Johannes said. His stories were like this—spare, and though they had good starts, they often went nowhere. "I brought home live fingerling trout in a soda bottle and stocked them in the mill creek near my home." Between the anecdotes he taught me words in Turkish: *ekmek, çakmak, tavuk.* But the most important were the words for trout, in Greek and Turkish, respectively, *pestrofa* and *alabalik,* and beer, *bira.* "*Bir* is one, so when you order a beer you say *bir bira.*"

"On the map the river is called the Vjosë," Johannes said as we looked into its emerald currents off a bridge. "That is the Albanian name. The Greeks call it the Voïdomátis."

We set up camp by an ancient stone bridge. Swallows darted above the surface of the water chasing insects. Below them, trout were rising to the same prey, small caddis flies. Johannes explained that this was a sanctuary closed to fishing, a national park, and for this reason there were many trout.

"We will wait a little before we fish, till the end of the day when the sun is lower and the hikers have gone home."

I sat in the shade of a giant plane tree and enjoyed the cool shade. Its trunk was mottled like a sycamore, and mimicked the play of light that reflected from the river on its broad leaves. The air smelled of honeysuckle.

"Ah, *mi hijo,*" my son, Ida said, coming up beside me to sit with her knitting. After a time, I left her there and walked to the river to wet my feet. The water was ice cold, fed either from melting snow in the mountains or from an underground spring. I knew that trout streams were usually cold—cold water held the high levels of dissolved oxygen that trout needed to survive.

When Johannes felt it was safe to fish, we found a quiet and hidden place to dress in our wet suits, beneath some ancient plane trees. With masks and snorkels, we dove into the river.

I had dived in tropical countries, over vast coral reefs in warm water with many-colored fishes. This was quite a different experience. Frigid water quickly seeped into my wet suit and formed a layer between the neoprene and my skin, eventually warmed by my body. We entered at the top of the pool and drifted quietly down like detritus, our eyes alert for any movement on the bottom. From above the surface you could see no trout; underneath now, I could see maybe fifty, or a hundred. They swarmed over the bald river stones like horseflies, scattering under logs and into dark crevices in the rock. Among the throngs were two giants, though I was mindful they looked bigger with the distortion of my mask and the water. The biggest fish took the best shelters, leaving the smaller fish vulnerable, like runts in a litter with no nipple. I could not help thinking how many trout I had failed to see in my life, staring into currents from above. This was a revelation.

The mouth of Johannes's net was a braid of wire coat hangers bent to form a circle. The mesh of the net was green and very fine, tapering down to a point, so that if a trout swam in it could not swim out. I watched Johannes use it, swimming after the fish, twirling and chasing with the swiftness of an otter, until he had trapped the fish under a rock and coaxed it to flee into the mesh. Though his element was air, I could not help but wonder if the amateur ichthyologist was part fish.

The brilliant colors of the trout were not apparent under water. It was in the light of the day that I saw the luminescent cobalt blue on the gill plate, the vermilion and black spots, irregular in shape.

We camped by the river in the national park that night after sharing a bottle of retsina wine.

The following morning we headed for Turkey, on a winding road up a mountain pass. The land was dry and punctuated by hard leafed oaks, changing to dense green foliage over the divide. Johannes turned onto a dirt road along a stream, a small tributary of the Aliakmon River, the location of the first known written reference to fishing with an artificial fly.

The closest there was to an ancient holy site for fly fishermen, the Aliakmon (which is thought to be the Astraeus mentioned below) and its Macedonian fly fishermen were mentioned in a second-century A.D. text called *De Natura Animalium* by Claudius Aelianus, a Roman naturalist:

> Between the cities of Beroea and Thessalonica flows a river called the Astraeus, and in this river are fish with spotted skins. These spotted fish feed on insects unique to this countryside, which flutter over the river. The insect is not like the flies found elsewhere, nor does it resemble the wasp in appearance, nor could one describe its configuration as like a midge or the bee, yet it had something of each of these. The river people call it Hippouros.

> The flies seek food above the river, but do not escape the attention of the spotted fish swimming below. When the fish observe a fly on the surface, they swim up stealthily, careful not to disturb the currents, lest they should frighten their prey and, coming upward like a shadow, open their mouths and seize the flies, like wolves carrying off sheep from the fold, or as eagles take geese from a farmyard—having captured the flies, the fish slip back into the rippling currents.
>
> Although the fishermen understand this, they cannot use these insects as bait, for when one touches them, they lose their natural coloring and their wings wither. The fish have nothing to do with such damaged flies and refuse them.
>
> But the fishermen have planned another snare for the spotted fishes, and deceived them with their craftiness. They wrap ruby-colored wool about their hooks, and wind about this wool two feathers, which grow under a cock's wattles and are the color of dark wax.
>
> Their rods are about six feet, with a line of a similar length attached. With this they cast their snare, and the fish, attracted and made foolish by the colors, come straight to take it.

I had tied several of these Hippouros flies—undoubtedly caddis flies, which I found to be very abundant in this drainage—and cast them into the stream we had stopped at near the town of Tripotamos. Intoxicated by the smell of sheep dung, I became uneasy and had trouble fishing. Neither Johannes nor I saw a trout.

We passed fields of blooming sunflowers, as determined to face the sun as Johannes was to travel east.

We came to the coast of the Aegean and the city of Kaválla. Palm trees grew on the beach, zinnias along the sidewalks and in front of resort hotels.

"*Deniz* is Turkish for sea," Johannes instructed at one moment, and then hearing cicadas singing from the roadside trees, he said, "I have found cicadas in trouts' stomachs."

We stopped at a café on the seaside. The waiter brought us three glasses of clear alcohol and three glasses of water. Johannes showed me that when you added water to it, it turned milky white. "It's ouzo," Johannes said. "They call it raki in Turkey or *aslan sütü,* lion's milk."

We set up camp that night by the sea.

It took two hours to pass through the border.

The official read my name aloud from my passport.

"James Otakar," he said, looking at me in the backseat. I had inherited Otakar from my grandfather. It was my middle name but appeared printed next to my first. We were on our way to Istanbul.

We came to the Marmara Sea and from the car I could see the broad-flanked ships sailing in it. It was Saturday and a long line of cars was headed the opposite way from us, out of Istanbul to the beach. By noon we were passing through the city, where Johannes refused to stop, and over the long bridge across the Bosporus Strait into Asia.

"Now we are in Asia," Johannes announced when we reached the other side.

"Too bad you can't see the city," Ida said, "the Hagia Sofia."

"But there are no trouts in Istanbul," Johannes said.

It was six hours to the lake where we would spend the night, Abant Gölü, and inquire among fishermen about the trout that lived there.

By late afternoon we were nearing our destination in the northeast, Abant Lake. Johannes handed me something to read that he had been keeping in the glove box. It was a scientific paper, our primary source for the locations of the trout in Turkey. I pulled it out of the plastic sleeve he had kept it in. The title page read:

*The Trouts of Asiatic Turkey*
The Hydrobiological Research Institute
University of Istanbul, April 1954
by Enrico Tortonese

As I read more, I learned that Tortonese was an Italian biologist from the University of Torino who had made several expeditions during the late 1940s in search of trout in Turkey. The paper outlined his observations and discoveries concerning the specimens he'd found, using the trout nearer his home, from Sardinia and Corsica, as benchmarks for comparison.

Tortonese was first to formally describe the trout of Abant Lake as *Salmo trutta abanticus.* The tone of the paper was formal and scientific, but I enjoyed the spare prose, which I found to be jewel-like and spirited.

> 1. *Uludåğ* (Blythinia)
>
> Locality—While on excursion on the mountains South of Bursa, I was able to examine several trouts just captured in cool brooks along the Northern slope of Uludåğ (Olympus); height about 2000m.

As if in some dream where at every turn there is a coincidental meeting, we arrived on the broad shores of Abant Lake the moment at which I came to observations of its trout in Tortonese's paper.

> The trouts living in Abant Lake form a small, isolated population. These fishes remind one of those present in some rivers of Sardinia, described by Pomini (1940). It can hardly be doubted that the trouts existing in Abant are genotypically as well as phenotypically distinct from those of Sardinia. What is their taxonomic status? Of course, there is no question about their belonging to *Salmo trutta,* but I do not think they are to be included either in *S. t. macrostigma* (the Mediterranean trout) or in *S. t. labrax* (the Black Sea trout). They rather represent an undescribed subspecies, for which the name *abanticus* is here proposed.

*Pierre Affre with my* silure, *Ile Saint-Louis, Paris.*

And as we took to a dirt road around the lake, set in a bowl-shaped valley surrounded by green mountains, we saw some fishermen casting their lines. Just as we reached them one of the fishermen was landing a fish. Like weary treasure hunters who had just glimpsed El Dorado, we jumped out of the car and descended on it. It was a golden trout with leopardlike spots that seemed to glow as it reflected the last light of the setting sun.

When we had finished admiring the fish, and the tall Turkish angler had stowed his prize in a canvas sack, we had an Efes beer and some dinner at a small wooden lodge by the lake. During the day, Johannes had been fairly silent, though I could not determine if that was due to his nature or his reluctance to speak Spanish. With alcohol now, and one trout under our belts, conversation came easier.

We all laughed and spoke in combinations of the languages we knew. We noticed each other's idiosyncracies. Ida discovered that I ate very quickly and in large quantities and joked that I had four stomachs like a cow. She mocked Johannes's tendency to take his mustache between his thumb and forefinger when in thought. Johannes joked about his own bald head, saying it made him faster as he chased trout in the water. Ida asked if I had a girlfriend, and when I told them about Yannid, she teased me, suggesting that I was a heartbreaker.

"To our first *alabalik,*" Johannes said, toasting, and took his mustache between his index finger and thumb. We laughed. "In Turkish," he said, "*balik* is fish and in Arabic *Allah* is God, so maybe trout is God's fish."

*Ala* in Turkish actually means a scarlet shade of red; it also means speckled, so *alabalik* is a speckled fish.

We had a glass of raki and that helped me to sleep. The air was crisp and fragrant by the lake that night.

## The Fig Eaters

Turkey was all one time zone and the farther east you traveled the earlier the sun rose. Near the border with Iran, Johannes said the sun was up at 4 A.M.

We were all a bit hungover the next morning, but still we joked and played etymological games. When we stopped for breakfast along the road headed east, Johannes wanted to order eggs but could not remember the word in Turkish. So when the waiter came, Johannes said the word for chicken, *tavuk,* and mimed the shape of an egg with his hands. This made Ida laugh and reminded her of a time from when she and Johannes had lived in Colombia.

"My aunt came to visit and went to a café to get a coffee with milk, *café con leche,* but didn't know the word for milk, so when the waiter came she said *café con moooo,* and squeezed her own tit." We laughed heartily.

When Johannes got up to go to the bathroom, Ida said to me, "He's probably going to look for trout in the toilet. A special trout." But she did not laugh.

That day we drove from the Black Sea basin, which had been lush and green, over the mountains to a more arid land in central Turkey. By the time we reached the capital of Turkey, Ankara, we were ready for some tea and lunch.

*Çay* was poured from a silver pot into small clear glasses, pieces of dark tea leaves visible on the bottom. The body of the glass was hot, so it was customarily held by the rim and sometimes drunk with a sugar cube in your mouth. We ate a fresh tomato salad with onion, pepper, cucumber, and parsley.

"The farther east you go, the less alcohol there is, and the more parsley."

After the yogurt soup we left for the city of Erzincan.

Somewhere, just east of central Turkey, we passed an old man driving a horse cart. We crossed a small bridge over a stream, which Johannes said was a tributary of the Euphrates. Tall rows of poplar trees grew along the banks. Their tops were tossed and their leaves shook by a warm and persistent dusty breeze.

Beyond another small stream on a dry hill I spotted an army-green tank with its large barrel pointing at the road. Soon, beyond the waves of heat rising from the road we could see men standing, and later their vehicles materialized beside them. They stopped us in the middle of the flat land.

"Passports," they said. They took and examined them. I examined Johannes's face in the rearview mirror to get some indication of the seriousness of this meeting. He looked calm and even had an awkward smile on his face.

"What is your purpose for traveling here?" they asked him.

"*Alabalik,*" he said.

The soldiers, who looked to be younger than me, probably eighteen or nineteen with barely any stubble on their faces, laughed and laughed, repeating the word *alabalik* over and over. They paused for a while and then said, "Cigarettes?"

Johannes handed the young soldiers a pack with two cigarettes left in it. They took them out, and stood expectantly. "Okay," Johannes said, and wrestled an unopened pack out of his shirt pocket. He handed it to them and they let us pass.

Farther down the road we passed more tanks with their barrels pointed at the road. It was disconcerting. We had entered an area of Turkey under emergency rule by the Turkish military.

For many years the Kurdish people who lived in eastern Turkey and considered themselves of a separate race and religion were trying to secede from Turkey and create their own nation of Kurdistan. To prevent the separatist and sometimes terrorist group called the PKK, Partiya Karkaren Kurdistan, Kurdistan Workers' Party, from achieving its goal of independence, the government had stationed troops there. Frequent outbreaks of fighting occurred, especially in the southeast, our eventual destination near the Iraqi border, where the headwater tributaries of the Tigris River flowed from craggy mountains. I knew from what I'd read that the Turkish military had a history of suspicion of outsiders; the added political turbulence would not make traveling in the region any easier.

I looked to Johannes to be my Virgil in the inferno and hoped that he was capable as a guide. The roads were unpaved and the fields beside the road were unevenly cut by hand with scythes.

After eight hundred kilometers on the road that day, we spent the night in Erzincan at the Hotel Berlin. It was a run-down establishment with no furniture in the lobby, just a desk and a telephone. We were required to leave our passports at reception overnight.

"I don't like the idea of leaving my passport here," I whispered to Johannes.

"You only have one passport?" Johannes asked, striking a match to light a cigarette.

"Don't tell me you have two?"

"Of course, so does Ida."

We rented one room with two beds and I slept on the floor. At about one in the morning I woke up with a dry mouth and a full bladder. There was no water to drink and I couldn't get the bathroom door open. Ida was snoring loudly, making gasping noises like a slain cow, and Johannes let out farts at intervals. I tried to fall back asleep but could not.

As the sun rose, and a cool breeze crept through the window, I heard the call of "God is good" broadcasted from the spires of every mosque in the city. It echoed through the street, a high plaintive song.

*"Allahuekbahhh, allahhhhalllaaahhh."*

The next morning I purchased several postcards depicting the mountains near Erzincan and wrote messages home on them as we drove east.

We crossed the Euphrates River twice as it wound back and forth under the road, and turned north to follow a smaller tributary called Balik Çay, or fish stream (Ç in Turkish is pronounced ch; *çay*, besides being the word for stream, is also tea in Turkish).

In the town of Mercan, at my request, Johannes stopped at a post office. At a table in front of the building, several mustachioed men were seated playing backgammon. They looked at us and squinted from the bright sun, chewing on dried figs they took from a pile between them.

*"Alabalik?"* Johannes asked.

They pointed in the direction we were going, away from the Firat Nehri (Turkish for the Euphrates), and said we would find trout in the village of Balikli, twenty kilometers distant. They offered us tea and figs but Johannes declined. "We don't have time for tea," he said to me, "we only have time for trout."

We drove toward a range of low snowcapped peaks, called the Otlukbeli, and arrived in the village the fig eaters spoke of. A small river flowed through Balikli, under a bridge and through a shady poplar grove. From the bridge we could see a skinny man, his long trousers rolled up, standing with a fishing pole in the middle of the stream. We stopped to watch him fish.

The water in the river was somewhat opaque and milky blue. It flowed around his legs and an eddy formed downstream of him where he dropped his bait, which looked to be a ball of bread.

Johannes called out to him over the sound of rushing water, *"Alabalik."*

"You think he's fishing for trout?" I asked Johannes.

It was nearly midday and the reflection of the sun on the rippled currents was blindingly bright. A cool breeze blew from the poplar grove carrying a strong scent of cow dung.

Our eyes were fixed on the fisherman when an old man came up behind us. He cleared his throat audibly to get our attention. When we turned, he lifted an open hand to his dark forehead as a kind of greeting.

*"Alabalik,"* he said after a long silence, *"evet,"* yes, and shook his head from side to side. Johannes looked at him with a hopeful glance. The old man spoke again like an echo of himself from parched lips under this thick gray mustache, *"Alabalik."*

When he said the word *alabalik*—the *i* is pronounced like the *u* in *put,* and the word was spoken deeply and aquatically—it was as if he were uttering a secret that should not be shouted. He seemed excited to help us. Perhaps, I thought, he was a fisherman himself.

Johannes wanted to make it clear that we were interested in the indigenous trout. He told him repeatedly, *"ala, ala,"* red, speckled, trying to communicate through mime and drawings in the dust on the hood of the Land Rover that we wanted only the native trout with the red spots.

*"Evet,"* the man said, and to show he understood he took out a small container of paprika from his pant pocket, spread some on the soft underside of his arm, and mixed the powder with a little spit to produce a soft vermilion color.

"Yes, yes!" Johannes shouted, delirious with pleasure.

The man pointed up the river. A wind rustled the leaves in the poplar grove.

*"Alabalik,"* he said, and looked at the sun and then his watch, "no," by which he meant to say the sun was too bright for fishing at that hour. *"Çay?"* he asked.

Johannes turned to me. "Sometimes we must drink tea to get trout," he said.

The old man told us to wait in the poplar grove while he prepared the tea. We sat beside a small aqueduct in the shade and waited for him. Ida brought her book to read, but put it away when the man returned carrying a tray of tall clear glasses and a silver pot.

He sat down with us, Indian style on the grass, and as he poured the tea, he began what I inferred, from the different pieces of languages he tried to use—German, French, and English—was a philosophical discussion about the virtues of country life over city life.

"Istanbul," he blurted, "no good," and spat in the dry earth. He took a bit of soil and rubbed it on his forehead, then filled his teacup with water from the aqueduct and drank it. "The sound of the stream is good for sleeping," he said through a combination of speech and mime, "the water is pure for drinking."

He told us his name was Celal Boz and shook our hands. Then he took out a pack of cigarettes from his shirt pocket, but there was only one left and the filter was broken off, so Johannes offered him one. He accepted, lit it, and puffed, staring into the foliage of the poplar trees above him. He took a sugar cube from a small bowl and pointed to cultivated fields on the other side of the river. "I farm sugar beets," he said in broken German. A small boy came up to Celal's tea tray and stole a lump of sugar. He put it in his mouth and sat within earshot of us. The tea had become stronger. We were nearing the leaves at the bottom of the pot, the sun was lower in the sky, and Johannes was impatient to go fishing.

But Celal was not impatient. His hospitality was thorough and unwavering. We would not go fishing until he gave us a tour of his home.

He carried the tea tray and we followed him to the steps of his abode, where we took off our shoes and walked in. The room was dim and colorful in contrast to the bright stark landscape outside. On the wall opposite the door were two portraits. One, Celal explained, was his father, and the other, Kemal Atatürk, the former president of Turkey (*Ata-türk,* or "father of the Turks," known for Westernizing the culture and introducing the Roman alphabet). Celal was proud to show us that he had a shower and a telephone. He offered us lodging for the night and invited us for dinner.

Johannes said the word again, *"alabalik,"* as a hint that he wanted to go.

True, Celal had agreed to be our guide, and remembered his original duty as soon as we stepped out into the cool dry air.

He sat in the passenger seat of the Land Rover as Johannes drove, pointing which way to go when we came to intersections in the dirt road. Where there were villages children came running from the homes, curious to see who was passing through. But there were few signs of people over the hour that we drove, just large uncultivated plains blooming with daisies, cosmos, and chicory.

At last we crossed a bridge over the stream again and Celal pointed below it and waved his hand. *"Alabalik çok,"* he said.

*"Çok?"* I said to Johannes, "what is that?"

"*Çok* is many," he said.

We stopped the car, and out of the streamside willows, it seemed, a dozen children ran up to us, curious to see what we were doing.

Ida read her book and attracted the least attention, especially I think because of her dark skin. The children were mildly interested in my fly rod, showing confusion in their faces as to what I intended to do with it. But what seemed to astound them the most, as well as Celal, was Johannes, who had stripped to his underwear and was dressing in a black rubber suit. When he jumped into the river below the bridge with a net, they stripped off their clothes and jumped in too.

The river was very swift, and the children were swept down in the current until they found footing on the gravel. I could see

Johannes was having trouble searching for trout; though the water was clear, there were no holes deep enough to dive in and the turbulence from the riffles obscured his view.

"I saw two," he called when he surfaced, and blew water out of his snorkel, "one big, one small."

The children gathered around, dripping water, waiting to see if the fisherman caught anything in his net. He hadn't. They put on their clothes and disappeared into the bushes again.

Ida and Celal agreed to stay behind with the car. The water looked fine for fly-fishing and I was eager to hike upstream. In five days of travel it was the first day I had taken out my fly rod. Johannes and I took to a path along the river and soon saw some men walking. We stopped to shake hands with them.

"Welcome," said the oldest man in English.

"*Alabalik,*" I said as a statement of purpose. He looked at my line.

"Not with this," he laughed. "It's too thick, the water is clear." He was looking at my green plastic fly line, not the thin clear end where the fly was tied. I showed him that the very end was clear and he looked at me and said that it was too thin, indicating with his hand that the current was strong. The man then looked at Johannes and his net.

"It is impossible to catch trout with this," he said. "You need a bigger net."

Johannes shook his head back and forth. "I have caught many trout with this," he said.

The men went on their way. A small way up the stream the path stopped and we were blazing our own trail through the grass and willows.

I began to fish with my fly rod. For some long while I had no luck and thought there were no trout. Then, as I was walking along the bank, I saw a dark shadow against the gravel. I looked up to see if it might have been the shadow of a passing bird, and seeing there were no birds I decided it must have been a trout.

I cast a weighted caddis fly into small pockets here and there up against the willows that grew on the bank. The current was swift and took the line before the fly could sink. I waded out in the fast water and held my rod over the calm pocket, dropping it in and letting it circle in the back eddy. Johannes was watching from the shade of an apple tree when I hooked a fish. It got in the main current and shot off downstream. I stood on the bank hopeless, imagining what the fish was and what it looked like, and the tippet broke.

"*Scheisser!*" Johannes yelled out with his arms in the air. "Why could you not bring the trout to the bank?"

"Me?" I cried. "Why couldn't you get in the water and net it?"

We were both disappointed because it had been a good-sized fish and neither of us had seen a fish from that stream.

We fished farther up the river and seemed to be getting closer to the mountains, but time was passing and I hadn't caught anything. "I'm going to head back," Johannes said, "maybe there are no more trout here."

"Maybe you are right," I said, "but I'm going to fish a bit more." Johannes left me there.

I walked a bit longer and came finally to a long and slow pool I had been looking for. I knew, from the streams I had seen and fished, that there would be a big fish in that pool. Green moss carpeted the stones near a small cascade at the head where a single arching willow branch concealed a very likely looking spot.

I stepped into the stream and did my best to cast to the top of the pool, stripping more and more fly line off my reel to extend my reach. My small caddis fly hit the water and drifted into a small whirlpool, where I let it sit for some time. When I finally lifted my rod I could not believe the heavy weight at the end. It was a good trout and I was so afraid to lose it that I began to sweat.

I fought it gingerly but kept pressure on the fish and my rod tip high. The fish jumped, and when it did I saw how golden it was and that it was a good fish, maybe two pounds.

When I had pulled the trout close to my feet I saw it against the colored gravel; its back was a deep olive-oil green. Its sides were golden, like the sun when it nears the horizon, and its lateral line was strung with a row of vermilion spots with blue-white halos.

Johannes was somewhere downstream of me if he had not already made it to the car. My first thought was to photograph the fish and let it go. I was angry at Johannes for turning back and didn't think he deserved to see it. I also did not want to kill the fish, but I felt I would regret not showing him what I had caught.

I reluctantly killed it by sharply hitting the top of its skull with a river stone. Then I returned with the fish, strung through the gill on a willow branch.

Back at the Land Rover, Johannes unfolded his examination table and took out some paper, a pencil, and a camera, and lay those items on the table.

"Good work," he said when I handed him the trout, "too bad you could not have brought him back alive." He measured the eye, the head, the fins, the distances between them, and counted the fin rays and the number of lateral scales, the pyloric ceca (small fingerlike pouches in the digestive tract), and gill rakers, recording all these quantitative observations on the page. Then he drew the fish, as I did, noting all its characteristics, and when he had finished, he cut out the liver and put it in a vial of alcohol, writing *Balik Çay, Euphrates basin, Salmo trutta* and the date on the label.

The tissue samples from this fish and any others we caught would be sent to Louis Bernatchez, the research biologist at Laval University in Quebec who was using DNA to create an evolutionary map of the trout of Europe and Asia (and to define the characteristics that make each lineage unique). Specimens of particularly interesting fish, Johannes said, we would preserve whole and send to Dr. Robert Behnke at Colorado State, the man who had introduced us.

"Send him a piece of your liver," Ida joked, observing Johannes's scientific rituals. "See if he can tell the difference." In the meantime, Celal Boz showed excitement over the nice trout, but expressed his impatience to get home.

We sat on Celal's cement porch in the early evening drinking tea, awaiting the dinner his wife was preparing for us. Buffalo began returning from pasture over the bridge on the river, kicking up dust as they walked. Their lowing mingled with the sounds of flowing water. Ida found her way into the kitchen, which was in another building, and helped Celal's wife cook over an open fire.

We could not refuse the Turk's hospitality. As we waited for dinner, he insisted we sleep on his mattresses and not the ground. Near twilight outside his front door in the poplar grove we heard the braying of donkeys, bells, flutes, and singing. A band of Gypsies had decided to take up residence there for the night and began pitching their tall canvas tents on the smooth green grass where we had taken tea that afternoon and where I had pitched my tent and intended to sleep. They lit two small campfires, and three Gypsy children lit sticks on fire and began racing around my tent. Celal offered again that we sleep in his home.

Finally Celal's wife and Ida, after two hours in the kitchen, brought out the trout I had caught, grilled over the fire, the skin crispy and golden and sprinkled with paprika. We ate it with flat bread and dishes with eggplant and yogurt. The door to Celal's home was left open so you could hear the sound of the river. As the sky grew dark, all you could see were two Gypsy campfires and the faces of the Gypsy children glowing in the warm light.

# The Trout of Eden—Cool Mountain Brooks of Gülyurt and Ararat

To explain Johannes's route to our eventual destination, southeast Turkey, which for several more days was circuitous and dizzying, one needed only to look on a map, understand our quarry, and realize the manner of roads relative to topography. There were still other trout he wanted to catch before we got to the upper tributaries of the Tigris River near the border with Iraq. Trout required cold water, which generally meant that they lived in streams high in the mountains, where the roads were poorest and most difficult to maintain.

The next morning we traveled north again and crossed from the Euphrates River watershed, which flows to the Indian Ocean, over a mountain pass back into the basin of the Black Sea. Near the divide, we made our way through a cool penetrating fog. Emerging from the low clouds we glimpsed small villages built of roughly cut stone. Manure, beaten flat and cut in bricks, was stacked in neat piles outside the doors. Even now in summer, it was burned for heat and cooking in the absence of trees (which I assumed had all been cut). The pungent smoke awakened my nostrils. In contrast to the dry land we had traveled the day before, this country was green. Tea and corn grew on terraced hillsides. Where the land was uncultivated, wildflowers were dense, lush, and in all colors imaginable. With the car windows open we could hear, more than we could see, in the mist, a small brook tumbling down the hillside.

At a dramatic viewpoint in the valley the mist had partially lifted, revealing a small town and a mosque that forced its tall needlelike tower into the low clouds. On the adjacent slope a woman was cutting grass with a scythe. Another woman was raking the grasses into a pile and a third was carrying a load down the hill by a tumpline about her head, her torso bent almost at a right angle to her legs.

We stopped on the road by a small brook and considered fishing. I decided to string up my fly rod and see if I could catch a trout. Out of the mist appeared three young men with the beginnings of mustaches on their faces. One approached me and removed a small golden hook from the leaves of his wallet and began to tie it on the end of my line. He encouraged me to follow him down the hill. Johannes said that I should, so I and the young man with the golden hook were running down a hill into the thickly green valley, through the damp cool air, wildflowers growing to our shoulders.

The stream was swift and clear and the rocks on the bank glistened with moisture. The young man turned over some river stones and strung my hook with live mayfly nymphs he found there. Then he took my rod and began to fish with it and I saw how agile he was. He leapt from wet rock to wet rock, dunking the nymphs in a series of emerald pools, almost a nymph or sprite himself. When he walked on the bank he barely left a mark on the mats of purple wildflowers.

He caught several trout the size of large sardines, heaving each into the air and over his head, landing them in the soft grass behind him. They came unhooked in the grass and I searched for them, usually listening for them rustling before I saw them.

I was looking for one, and trying to listen over the sound of the rushing brook, when I looked up and saw my young fisher sprite had disappeared. The mist had become dense where he stood. I was just about to tell him to stop fishing, that four trout were more than I wanted to see killed, and then he was gone.

I stood up and began to run frantically downstream. Had he slipped and fallen into a deep hole, where he drowned or froze in the chill water, or worse, had the nimble nymph tricked me, secretly planned to steal my fishing rod, and was hiding with it in the tall orange and purple flowers? I felt confused and misled; I slipped on the stones as I ran after him, fell, and banged my knee, but I was determined to get my rod back. I was just as at home on the stream as he, after all, this was my domain too.

I came to a rickety wooden bridge and decided not to cross, but ran deeper down the valley and into the mist. I was picturing the young fisher continuing to fish down the stream with his golden hook and my rod, catching trout and heaving them over his head, when there ahead of me on a gravel bar I saw my rod and reel. I didn't notice until then that I was out of breath, and though the air was cool I was sweating. I reached down to grab my rod, so relieved to see it, as it was the only one I had brought on the trip. I lifted it up and inspected it; I could see nothing wrong and the golden hook was still on the line.

I looked around and saw no trace of the young man, but there on the gravel bar now I saw a girl was washing clothes. She was beautiful and dark skinned, wore a red cloth over the top of her head, and didn't appear to be startled by my abrupt appearance. I nodded to her; perhaps she had startled the sprite and caused him to drop the rod. Then I turned and began walking back up the valley to the road.

Johannes, Ida, and I spent that night on the Black Sea in Hopa, where Georgian fishermen docked their boats. We talked over dinner about the fish I had brought back that day. Ida spoke of all the types of wildflowers she had seen and photographed. Johannes did not listen or respond to her; they sat side by side but he did not touch her. I thought of all the wildflowers I had trampled while in pursuit of the young fisherman down the valley. Ida conceded that trout could at times be as beautiful to her as the flowers and kissed Johannes on the cheek. When she turned to light a cigarette I noticed Johannes wiping his face where her lips had touched him.

The next morning we drove over another mountain pass into a

third drainage, that of the Caspian Sea. On the pass the wind was strong and ripped from the ridge down the valley. We stopped by a small stream and asked an old man and woman if there were trout. A cold drizzle was falling.

"*Alabalik?*" Johannes said with his best accent.

"*Yok,*" the man uttered, nodding slowly and without an expression, none.

We drove into a lightning storm on a broad treeless plain. Over a small hill, through the thick fog and rain that was becoming harder and steadier, we saw Çildir Lake, and no boats or habitations anywhere.

Ida had taken charge of navigation and wore a look of concentration on her face as she studied the map. We passed a little sign by the vast lake that read Balik Restaurant.

"Pull over, Hannes," she demanded. "Didn't you think that James and I might be hungry?"

"How about me?" he said.

"You wouldn't eat if we didn't tell you it was time," she shouted, folding her arms over her chubby figure.

The bumpy road followed a peninsula out into the lake. We passed sod-roofed homes, mounds in the earth with little chimneys coming out of them, puffing smoke. We parked the car and walked into the small shack that we assumed was the fish restaurant.

When we were inside the restaurant, it started to rain even harder. There were four tables, and at one were several grubby-looking men with bushy mustaches. We sat down too; there were no menus, so we asked for beer and hoped the food would come. A boy brought us each a beer and a tomato and cucumber salad with so much parsley I could barely see the vegetables.

"See what I told you," Johannes said, "the farther east you go the earlier the sun rises and the more parsley on the salad."

*The* Schwarzfischer.

We drank our beers and then drank a second round. The brand of beer was Efes and we were surprised they had any at all, being that the area was mostly Muslim. We tried to guess what kind of fish they would bring us for dinner.

"Maybe they will serve us trout from the lake," Johannes said. The boy returned with three fish, butterfly filleted, cooked in salt, pepper, and paprika to a rust brown color.

"The fish is very good," said Johannes, chewing, and taking some more with his fork. "It would be even better with a little wine. But it is not a trout. My guess is that it is an introduced *Corregonus* species. You call it whitefish."

An old lady, hunched over and walking with the use of a cane, came in from the rain and sat down. I noticed she was staring at Ida. Ida noticed the staring too. The old woman came and sat down next to her, talking loudly in her ear as if that would make her understand better. The old woman smiled at her, with a mouth like a wrinkled pumpkin. I smiled at the old woman, but her eyes were fixed on Ida. She seemed to be curious about her, asking her where she came from, perhaps because Ida's skin was dark and she was traveling with us and our skin was white. "Is this your son?" she said, pointing at me. One of the men in the room spoke German and translated. She insisted Ida was Turkish, but Ida insisted she was from Austria, that she was dark because her parents were Indian.

After a few minutes, Johannes started speaking in German with the man who had translated for Ida and asked him if we could park beside the shack for the night and sleep. It was still raining and the man said we'd better not sleep on the ground in the rain. One offered us a room of his house to stay in.

It was too early to sleep so we kept drinking. We listened to the thunder and watched the lightning flash outside until only the old woman and the young boy were left. The boy put on some Turkish music. Then the power went out and the music with it. The boy went over to the radio and tapped on it with his hand but it did not respond. The old woman began to hum a tune.

It was very cold and misty by the lake the next morning. We bundled up with what warm clothes we had and drove around the other side. At the village of Gölbelen a man told us there were trout in a stream three kilometers up the road. But from that point beyond the village the road was impassable, even for the Land Rover. Johannes and I started off on foot, wiping away the sleep from our eyes, and Ida stayed behind with the car. Purple bell-shaped flowers hung above the dew-dampened grass. A man on horseback carrying a large scythe passed us on the road.

We walked up two kilometers and found the stream. Immediately, I began to catch trout on my fly rod. Since the trout were in a stream that flowed to the Caspian Sea, they would most likely be known as the Caspian Sea subspecies, *Salmo trutta caspius.* They were red-and-black-spotted like the Euphrates trout and the lavender spot on the gill plate was the color of the bell-shaped flowers that grew in the rich green meadow.

Two or three pools upstream of me in the tall grass I saw a man fishing the stream with a cast net. He stood on the bank and threw the net into the small pools. As he did, the meshes glistened like a cobweb hung with dewdrops. Johannes and I began walking back to the car with two fish kept alive in a plastic bag full of water.

"Today we will see Mount Ararat," Johannes announced as we were returning over the largely bare, windswept country. "Do you know about Ararat?" he asked.

"No, I don't," I answered.

"It's the mountain on top of which Noah had landed his ark when the earth flooded. Eden was here too. There must have been trout there," he said, half joking, as he walked. "And they survived the flood, of course."

On the road south toward Mount Ararat, near to the borders of both Armenia and Iran, Ida asked if we could stop beside an outcropping of red-orange rock to buy peaches from a street vendor.

Beyond there Johannes spotted a tank perched on a hill. When we continued on, we were stopped by the military. The soldiers, armed with automatic weapons, asked us to step out of the car and they performed a thorough search.

"What are you doing here?" one soldier asked in English as we stood outside the car.

"We are fishing," I said, but he didn't understand.

"*Alabalik,*" Johannes said.

"*Alabalik? Hah!*" the soldier said, and started to laugh. He had a long look at Ida and then waved us on.

A broad-based mountain with a white cap materialized in the distance, a shade darker in value than the sky peaked with a white cap. It was Mount Ararat, and though it was still twenty kilometers distant, it was so massive that from where I was sitting in the backseat of the Land Rover it spanned the entire windshield.

The living cargo of Noah's ark had survived forty days and forty nights of rain and storm. When the waters ultimately receded, they disembarked on the 41st parallel.

> Ch. 8:4 And the ark rested in the seventh month, on the seventeenth day of the month, upon the mountains of Ararat. 5 And the waters decreased continually until the tenth month: in the tenth month, on the first day of the month, were the tops of the mountains seen. 6 And it came to pass at the end of forty days, that Noah opened the window of the ark which he had made: 7 And he sent forth a raven, which went forth to and fro, until the waters were dried up from the earth.

Trout, of course, as Johannes pointed out, being denizens of the water, were not dependent on the ark for survival. One could say that God favored fish and chose a fate for life on earth that they would be immune to.

Heading on a long straight road to the mountain's base and another military checkpoint, we were also nearing Eden. The soldiers themselves might be guarding the gates.

In Milton's *Paradise Lost,* a spring bubbled forth near the tree of life in the Garden of Eden. That spring was the source of the Tigris River, and if we could find it, Johannes was betting there would be trout.

> There was a place,
> Now not, though sin, not time, first wrought the change,
> Where Tigris at the foot of Paradise
> Into a gulf shot under ground, till part
> Rose up a fountain by the Tree of Life (IX: 69)

In my mind, the sources of all rivers were good to drink from and to fish in: pure, cold, and clear. The walls of Eden had crumbled long ago but the trout were probably still there. Johannes was the only man I knew who was convinced that trout really did live at the source of the Tigris, and the closer we got there the more apparent became his monomania for catching one. Here there were no flowers except what bloomed from the chicory and sage. Fine dust filled the Land Rover as it blew across the land, settling on every horizontal surface, including my eyelids and upper lip.

At dinner that night in Dogubayazit on the Iranian border, Johannes discussed our journey's most risky task—to enter the region south of Lake Van, where the most hostile battles between the Turkish military and the Kurdish people were being fought, and emerge with a specimen of Tigris trout.

"You must now add the Kurdish word for trout to your vocabulary," Johannes said. *"Massi alé."*

The next morning we were even closer, on a lake south of Ararat described as having trout in Tortonese's paper. Balik Gölü, it was called. A sparse population of Kurdish nomads had pitched their brown tents over piles of stones on the lakeside. Where there was a road, the road was dry and hard and dust came in through the car's ventilation system. Johannes stopped the car by a boat beached on the shore. Beside it a fisherman was sleeping in his car.

*"Alabalik?"* Johannes said, knocking on the man's window. The man got out of the car, brushed some dandruff off the lapels of his gray suit, and offered two trout he had caught that morning in his nets. The fish were mangled from fighting in the fisherman's net, but you could still see the characteristics that convinced Tortonese of the fish's uniqueness: the sparse black spotting pattern, the lean sleek body, and the forked tail.

We came to a town called Diyadin. Above the village there was a stream called Murat Çay that Johannes wished to sample for trout. He stopped at a kabob stand on the main street where men were gathered and tried to recruit one or two to be our guide. He asked them if there were trout in Murat Çay, holding up the trout we had bought at Balik Lake to demonstrate.

"There are many," one man told us. "But you shouldn't go because there is fighting in the hills."

*"Beş milyon, bir alabalik,"* Johannes said, offering five million Turkish lirasi for one trout. None of the men would take payment to guide us up the river.

"They shoot you in the head," they warned, demonstrating with their fingers.

We gave the two trout to the cook at the kabob stand and asked if he would cook them. He cooked them well; Ida ate the smaller of the two and Johannes and I split the big one. Johannes was picking his teeth with the spines when a man invited us into his office for tea.

The man had a large kind smile and served us tea in red lacquered cups. Judging by the tools in his office and the tooth diagram on the wall, I surmised the man was a dentist. There was a map of Turkey on his desk and Johannes told him where we wanted to go. The man showed disapproval.

"Don't go up Murat Çay," he insisted, "there are terrorists in the hills."

After we'd finished our tea he sat Johannes in the dentist chair for a free inspection. He told him to open his mouth.

"Ha, ha, ha," the dentist said, looking into Johannes's mouth. Then the dentist made an imaginary gun with his hand and pointed the barrel at Johannes's head. Ida laughed and took some photos, the flash filling the dark office with light. The warnings did not seem to scare her.

"Don't go to Murat Çay," he repeated, "ta ta ta ta ta . . . ," and then pointed the imaginary gun at Ida's heart, "Ta ta ta ta ta. . . ." The dentist laughed.

*"Qué dentista,"* Ida said, rolling her eyes. *"El es loco."*

Johannes jumped out of the dentist chair and we left the office. The dentist stood smiling and laughing in the doorway as we walked away.

I noticed for the first time as we were leaving the village of Diyadin that white roots were showing in Ida's black hair. It had not occurred to me that she was old enough to dye her hair, she had such a youthful and jovial demeanor. The white roots reminded me that time had passed since we'd left Sankt Veit.

The checkpoints became more numerous as we drove toward the city of Van and the vast turquoise lake of the same name. As my anxiety grew, I could only imagine what was going through Johannes's head. He was probably thinking he'd discover a new species of trout. I was thinking of the dentist.

By noon we were deep into southeast Turkey and finally in the drainage of the Dicle Nehri, as the Turks call the Tigris River. Everywhere on the road and beside us across the semidesert countryside were reminders that a war was being fought here. There were tanks, armed vehicles, men with guns, blown-out and barricaded sections of road. As we approached an unusually large checkpoint, Johannes made sure to remind me not to take photos of the military, to hide my camera and my journals.

The road was blocked by two iron gates and two soldiers waved us to the side of the road. Johannes stopped and opened the window.

"What is your purpose for traveling to the village of Çatak?" the soldier asked in English.

"*Alabalik,*" Johannes said.

"*Alabalik,*" the soldier repeated and sunk his upper lip into his lower one and wrinkled his brow. "Well, passports, please."

He held our passports together in his left hand and licked his finger in anticipation of flipping through, but he did not open them.

"Please step out of the car."

What might they find that would cause suspicion, I wondered. Would they confiscate my sketchbook, my journal, the film I had shot? As we sat in the dirt and waited, I thought about how the soldiers looked to be about my age. Johannes, Ida, and I sat on the side of the road. Eventually the official who held our passports returned.

"Do you really want to cross this line?" He looked at Johannes. "If you do you may, but the road to Çatak will be closed after 5 P.M. At that hour we are sending in a military operation, and you should not be there."

The official dusted off his hands and lifted one of the metal gates to let us pass. It was 3:15.

"We have only an hour to fish," Johannes said, driving desperately fast down the only road that led to the streams near the Tigris's source.

Ida in the meantime stared out the window until she began to shout at Johannes. I could not make out exactly what she was saying but she involved me in her tirade and she was very angry.

"You should have left me at home. I would have stayed there had I known you'd put our lives at risk."

Sweat began to bead up on Johannes's brow and dripped down to his forehead and the bridge of his nose, hanging as a drop on the tip, where he wiped it off. I took my mind off the shouting by concentrating on the map. I saw that the town of Çatak was only about forty kilometers from the border with Iraq. Between us and that border, a dozen or so streams trickled out of the mountains and gathered to feed the Tigris River. The stream we were headed for was called Çatak Çay.

Not twenty minutes had passed before the stream came into view, flowing through farms alongside the road. Large poplar trees shaded the banks, beans grew on poles, and squash and cucumber blossomed in rows in the dark earth. The stream itself was swift and clear.

"There must be trout in it," Johannes said.

But he did not stop to fish it, he drove all the way to the village of Çatak to ask the native Kurdish people where the trout were, using the Kurdish word for trout, *massi alé*. When we got there, though, Çatak was not a friendly scene. The streets were deserted and an armed soldier stood on every street corner. Johannes turned around and we headed back north of town.

Johannes pulled off the road near a lovely stretch of river. I was stringing up my fly rod when a red van pulled off the road beside us. Out of it poured a dozen Turkish police in plain clothes.

"*Alabalik?*" Johannes asked them.

"*Evet,*" they said, "passports?"

"You must leave," another said.

"Why?" asked Johannes.

"There is terror here, it is not safe," he said matter-of-factly, pointing up in the dry craggy hills overhead. "Thousands of people die fighting here."

I couldn't associate any permutation of the word "terror" with this scene. We were in a canyon of sorts and along the stream grew willows and poplar, their gentle leaves sensitive to the slightest breeze. The water was very cold, the sun burned my neck. I was fly-fishing.

"Okay then," the policeman said "you have ten minutes to fish, then you must leave." A colleague of his took a fishing rod out of the van on which was tied a large copper spoon. Another held a revolver, and I pretended not to notice. I was very thirsty and walked down near the stream to drink where a spring cascaded out of the rock on the bank. I drank until my mouth was numb from the cold water. I closed my eyes, then looked into the sun, thankful for such a beautiful day, then closed my eyes again and felt the sun burn from above the cliffs.

I was crouched down, balanced on my toes, when the ground began to shake and a raucous sound drowned out the gurgling of the small spring and the river. A tank was passing on the road.

One of the men approached and tapped me on the shoulder. "You must leave," he said. I had not made even one cast.

I counted ninety-three tanks passing on the road as we left the area, each with a soldier standing upright and armed in the open hatch. Kurdish children along the side of the road waved to the Turkish soldiers and some soldiers waved back. We followed the road back to the checkpoint and spent the night by turquoise-blue Lake Van under a large apricot tree.

## Returning to Austria

The next morning was the twenty-sixth of July. During nervous fits in my tent the night previous, I dreamt that I was back together with my high school girlfriend, Whitney. I would not have recognized her were it not for the freckles and her blue eyes and the way she wore her straight brown hair. With the dream came old anxieties. I could smell the cool melancholy air of the New England autumn when school began and the dreaded end of summer with it. An apricot leaf, turned yellow, had fallen into my lap as I read in the vestibule of my tent that morning.

Meanwhile, Johannes was having his own anxieties. He realized he was not going to catch his Tigris trout on this trip. I was picking apricots from the tree pretending not to hear the jokes he cracked about trading me to terrorists for a specimen of the Tigris trout.

During the course of the day's driving we came to more military checkpoints. The inspections were overwhelming, invasive, and taxing. For the first time Johannes had truly given up on his mission and we were headed west.

I pitched my tent that night on a rocky surface in the parking lot of a small restaurant in Malatya. I had bad stomach pains, perhaps from eating too many apricots, and the air was stifling hot.

The next day I lost my appetite, acquired severe nausea, and purged my intestines. I had two bites of bread for breakfast with my morning tea but could eat no more. My mind could concentrate on little else but the pain I was feeling. I suddenly felt dirty and couldn't remember the last time I had showered. Had it been eight days? I noticed for the first time that my arms were caked with dust. The thought of being clean again began to obsess me. I dreamed of having a shower and cold clean tap water to drink. What thoughts filled my head; perfect images of the pink faucet in the bathroom at home and how in the summer it beaded with condensation when cold water coursed through it.

I would jump into the next river that Johannes had in mind to search for trout. That happened to be the Karagöz River, a tributary of the Zamante, near the village of Pinarbaşi in south central Turkey. We had long left the basin of the Tigris River; the Karagöz flowed from its source to the Mediterranean Sea.

The Karagöz was Edenic; an even-flowing stream that wound lazily through cow pastures and meadows of tall grass where boys slept on golden mounds of hay. We stood beside the banks and watched the smooth clear river glide by and the tall green grasses show their silver undersides to the wind. It was a strong antidote for sickness and the anxiety the military presence had caused. I jumped into the water and scrubbed myself, then I lay down near the bank and fell asleep.

While I rested, Johannes dove in the water with his mask and snorkel. He explained when I woke that this stream held a very special kind of trout, a subgenus of the brown trout called *Platysalmo platycephalus.*

"Six years ago I fished here and caught many," Johannes said, "today I have not seen a single one. The water is not as clear as I remembered it and algae is growing on the bottom." We drove upstream to see if the water was clearer near its source and there we witnessed the problem. They had mined something from the hillside, and silt from the operation had flowed into the stream. The gravel necessary for the trouts' spawning had been covered with a fine gray mud.

"It is possible that this unique trout is now extinct," Johannes said as he stared in the water. "War may be the only hope for them."

Visibility decreased toward twilight as we drove through sheets of dust, backlit by a warm orange sunlight.

Late the next morning we were already on the west coast of Turkey, driving through a dense hardwood forest at the foot of Mount Uludåğ (Olympus). We stopped for a cucumber salad and yogurt with a bottle of Coke near the town of Bursa on the Marmara Sea.

As we approached the town of Edremit on the Mediterranean coast that evening I wrote in my journal of my surprise at the diverse nature of the trout we had encountered on the trip. Every population we had seen looked slightly different, and I imagined the differences were not only physical but genetic.

Johannes and Ida had arranged for us to stay with an Austrian friend named Ekhart who lived with his Turkish wife in Akçay, a clean white-plaster town with palm trees on the Mediterranean. We found their apartment building on a quiet street. Ekhart greeted us at the door and gave both Ida and Johannes a full embrace before leading us up the stairs to his apartment. As he spoke in German, I could understand only a little of what he said. He was concerned about Johannes's plan to return to Austria through former Yugoslavia, nations that had found recent peace after eight years of war. Ekhart sat us down in the kitchen.

"I read in the paper, Hannes," Ekhart warned, "that it is dangerous to pass through Serbia. The country is in chaos. Highways and borders are closed, bridges across major rivers are blown out. Corrupt policemen are stealing from travelers. The war is over, everyone is an opportunist."

Johannes listened. "It is a crowded ferry ride from Greece to Trieste," he responded. "I would rather drive and take the risk. Besides, there are many interesting trouts there."

Ekhart took Ida's hand in sympathy and kissed it. "I'm sorry," he said.

"It's okay," Ida said, "I'm used to it."

Shortly, Ekhart's wife appeared and greeted us, a lovely Turkish woman who spoke perfect German. She poured us each a glass of sweet wine like a muscat. It was cool and foreign, and seemed to plane the dust off the inside of my mouth.

Ekhart talked for some time as we sat around the kitchen table. I just kept filling my glass with muscat and watching Johannes, who seemed to be as indifferent to what Ekhart had to say as I was. Ida told them about our trip so far, which she described as a "horror trip." Johannes smiled occasionally and lit a cigarette, which appeared to shield him as he smoked it. His eyes were focused on something beyond the table where we sat. Ekhart and his lovely wife held hands as they spoke.

That night the five of us ate dinner at a restaurant near the beach. A truck drove around spraying insecticides, the mist settling in my drink and on my food. In an uncharacteristically autobiographical moment (he was drunk), Johannes told stories of his carefree twenties and how he first came to dive for trout. I smelled a warm salt breeze and heard the crashing waves.

"I used to drive from Austria to the Croatian coast and go skin diving. I became interested in the sea fish and began catching them for my tank at home in a small net I had made, not much different from the net I use now. On the return from one trip I passed by a river with very clear water and I saw fish in it so I decided to dive. It was the Krka River, and I didn't know how cold the water would be, or why it was so cold, that it came from big springs underground. I also didn't know the fish were trout, but when I dove I caught one and guessed that's what it was."

Johannes poured me a glass of raki from a carafe. He spoke about his apprenticeship as a young baker in Colombia, and about the two years he spent in South Africa. He suggested that I should enjoy being young and free. I added some water to my glass of raki, drank a bit, and yawned.

*"Estas cansado?"* Johannes asked me, are you tired? And then he looked at Ida. *"Es mejor estar cansado que casado,"* he whispered,

which in Spanish was clever and funny because the words for tired and married were the same but for one letter. It is better to be tired than married, he had said.

We paid the bill and walked back to Ekhart's.

The next morning Johannes and I drove inland, up the valleys between the coastal hills, through winding roads in olive groves, along rows of tall cypress. We passed a young boy shouldering a shotgun, a pair of songbirds tied by the feet to the end of the barrel.

We fished high above the coastal towns, and could often glimpse the sea. The trunks of the olive trees were black, almost charred looking, and their leaves were thin and powdery green. The oldest olives had been cut back many times, their hollow and ghoulish trunks sending out new and fertile shoots. They flailed their entwined limbs, dancing in a wild ecstasy, it seemed, to some music in the wind.

The next morning we departed Ekhart's on a ferry from the Turkish coast to Greece. Camped beside the sea that night, I had a distinct and memorable dream involving Johannes.

In my dream Johannes and I had hiked up a mountain and stopped to eat our lunch at the summit, seated on the edge of a cliff. We were calm and civil and not afraid of being so close to the edge until we had a look beyond it, and saw that below us was a deep abyss. After I had looked over the edge I could not get off my hands and knees for fear I might fall. And then, watching from my position, I saw Johannes walk to the edge, spread his arms as if he were stretching, and then leap.

His action to do so was intentional; he even had a slight smirk on his face. And as he flew in the wind and my view followed his calm descent, I yelled, "Help me, help me."

I hiked down the mountain to look for a ranger's office and as I did I tried to decide whether I should call Ida and if I did what I should say. I thought it would be just as well for anyone else to tell her what happened. Maybe I'd write a detailed message of the event and deliver it to her.

As I was drafting the letter in my mind, I was thinking that Ida would surely give me Johannes's books, research materials, and photos on trout. Of course, it was the first thing she would want to get rid of, remnants of the obsession that consumed him and had diverted his cares and attentions from her. I wouldn't ask, but knew she would give the collection to me. I would tell Ida that Johannes's death had been an accident.

Then I walked to a bar and had a drink. I was seated beside a man, and then recognized that it was Johannes's ghost, standing, with one arm on the bar, as he always did. I didn't act surprised to see him, I only asked him why he had jumped. "When did you decide to do this?"

"Monday, while I was driving," he replied. I tried to think back to Monday and where we had been, but I could not remember the days and places.

That day we had passed all of Greece in a rainstorm and now were traveling west through dark tunnels to the border with Macedonia. The border station looked abandoned but an agent emerged as we approached, checked our passports, issued us transit visas, and let us through. After some driving through a green and mountainous country, we crossed another border into Serbia.

Some kilometers down the road we were halted by police at a barricade.

"Passports," one said. Another looked at our plates.

"Fucking Austrian!" he yelled in English and waved us on.

The countryside was beautiful and the new corn in the fields held no ancient hatreds, as some of the people did. We soon found our currencies were not accepted, our dollars, shillings, and lira not exchangeable.

In the late afternoon we began to look for a hotel where we could stay the night. The proprietors told us that the only valid currency was deutschmarks and we had none. A thunderstorm blew over the road on our way to Priština. As our money was no good we could not even buy gas, though the tank was still half-full. Toward twilight, rain still falling, we pulled off the road where we would not be seen and camped in a farmer's field.

It rained all the next morning as we broke camp and all the way into Montenegro. We were descending to the Mediterranean and Johannes conserved gas by keeping his foot off the accelerator as we coasted down hills. We stopped on the banks of the Morača River to eat some canned meats and a dried sausage that Ida had brought from Austria. It was a dramatic river valley with mountains on both banks, the river forming deep pools in narrow canyons. We joked there, eating and staring into the water, about how we had camped covertly in the field. For some reason it seemed to amuse us.

That afternoon Johannes drove us all the way to the coast to the town of Kotor, built on a well-protected harbor nestled in a bowl-shaped valley. It was a beautiful town and the kind of place you'd want to spend a week or two, but Johannes was determined to get out of Serbia and Montenegro. We were near a border crossing into Croatia but people in town told us it was blocked. Everything had changed, they said, since the war, they hadn't bothered to travel out of the country for several years.

We tried the crossing; it was near the town of Igalo. But the police there denied our entry into Croatia. So we drove on, to a small road that led to a crossing at the border with Bosnia. The border guards there were more amiable. They spoke English, so when they asked for money I was the one who negotiated how much we would give them in American dollars.

*Mosque near the Black Sea, Turkey.*

"Prosek," the man said, looking at my passport. He smiled; the name seemed to please him, and he repeated it. He took an interest in it. "Where is your name from?" he said.

"My family is Czech," I said.

"I was thinking you were Czech," he said, "though you know Prosek is an old Serbian word. It means 'cutting through,' like a river through a gorge. There are several river gorges I know called Prosek. Be careful on the road ahead," he said, stamping our passports. "Don't stray from the road too far, there are many mines."

Across the border there was a noticeable difference in the health of the roads and villages. Ahead of us an entire section of asphalt had been blown out and a charred tank sat there covered in graffiti. Had we not been in the Land Rover it would have been difficult to cross. In fields beside the road were numerous small signs with the word "mine" written on them in capital letters. "The threat of mines is serious," Johannes said, pointing out the signs to me. He and Ida looked as nervous as I felt.

Villages ahead of us were completely destroyed and where homes were not leveled their roofs had been blown off, making them uninhabitable. The land in places looked like it had been torched, the vestiges of tall trees were black and rows of bushes a sienna brown. Portions of stone walls had been blown out, holes blown clear through stone churches, power lines lay on the ground, house chimneys stood like monuments to the events that exposed them. Most likely these abandoned villages had been occupied by Muslims or other non-Serbs and were destroyed by the Serbian military, the inhabitants either killed, put in camps, or forced to go elsewhere.

Johannes looked ahead only and did not turn to see any of the devastation for a second time. I could not help occasionally turning my head.

As twilight grew closer our, gas supply emptied. Johannes seemed shaken, and was more eager than ever to find a border crossing into Croatia. I stared at him and tried to find comfort in his determined face and the unshaven reddish hair that had grown up around his mustache.

We planned to cross in the town of Metković, but getting there proved difficult. We came to blown-out bridges, and were forced to take small side roads that were not on our map, where grass grew through the asphalt. We passed a charred vineyard, the burned lattice standing like a field of crosses.

There were certain precautions Johannes had taken that I had taken for granted. I realized this when the sun set, and Johannes said, "I prefer not to drive at night." It was the first time on the trip that we were driving in the dark. A car passed us without headlights. "Idiot," Johannes yelled.

We had very little gas left but Johannes was optimistic that we would make the border. He told me that we were nearing territory that was familiar to him, a good trout river called the Buna, but he had not been there since the wars had begun in Yugoslavia eight years before. He stopped at a campground where he and Ida had once stayed, but the entrance was chained off, marked by a sign with a skull and crossbones and the word "mine."

Farther down the road and close to the Neretva River we saw lights flashing and heard loud music playing. Young people were drinking and dancing at an outdoor bar. They were the first people we saw enjoying themselves since we'd entered former Yugoslavia. Shortly after passing the bar we crossed the border into Croatia.

After a good sleep on a firm bed at a hotel in Metković, our currency and credit cards valid again, I felt refreshed.

Over an egg-and-ham breakfast we discussed our options for the next several days. Johannes had to return to work but we had at least two full days to fish. It was Friday, after all, and he might as well return to work at the beginning of the week, on a Monday.

We decided to return to Bosnia that morning. Johannes said I must see and fish a beautiful stream called the Buna that tumbled full force out of a cave. He and Ida knew the owner of a café near the source of the stream, but alas, when we arrived, the café had been destroyed. The trout, however, were abundant, more so, Johannes said, than the last time he had been there, before the war began.

The trout I caught there with my fly rod on small caddis larvae imitations were a peculiar species unique to the Balkans that Johannes called the softmouth trout, *Salmothymus obtusirostris.* Their mouths were smaller than the typical trout and the upper jaw extended over the lower, suggesting an evolutionary preference for feeding on the bottom. They were beautiful fish, slightly golden with red and black spots.

The most destruction we witnessed was on the next stretch of road, as we returned through the city of Mostar to the Croatian coast. Mostar was an old Yugoslav city on the Neretva River, downstream of where the Buna flowed into it. Not a building I could see had been left unmarked by artillery fire. Empty shells from all manner of fire littered the gutters in the streets. It was in complete ruin. Ancient and delicately arching stone bridges over the river had been smashed at their peaks. But here, more than in any village I had seen, the people were active, rebuilding, planting flowers, laughing and smiling as if they had not noticed what happened to their town. "It is remarkable," Ida observed, after days of relative silence.

Eight years previous, before the war had begun, Johannes and Ida made regular trips in the summer to a small inn on the Adriatic coast run by an elderly couple who spoke German. Johannes and Ida were happy when we arrived in the old fishing village of Seline to find the couple alive and still running the inn. Only the wife remembered them and her German. She explained that her husband had forgotten his German and most other things after years of hiding in the basement during the war.

Through a sense of duty, the old man sat Johannes, Ida, and me down at a table on a small terrace. I took deep breaths of the air, fragrant with sea smells. The wife brought us each a beer and a salad of fresh lettuce and tomatoes, she said, from her garden.

The old woman served us a whole grilled mackerel on a platter and some grilled calamari. Then she brought out several clear unlabeled bottles filled with grappa. Time passed, different bottles were drunk from and toasts made. We drank out of a common sense of relief; to be near home, to be in a place where we felt safe. I hadn't realized how much I'd drank until I stood up to use the head and dizzily tumbled back into my chair. Ida was drunk too and suddenly lashed out at Johannes.

"Bah, Hannes. You don't care about anyone! How many years of begging did it take before you bought Mariela a horse? And do you buy me anything? We don't even see each other. When you get home from work I leave for work. You don't even sleep at the same time I do." She lit a cigarette and waved it at him. "You are here with me now, but you're not here." She waved her cigarette again to get his attention. "Where are you?"

Life was not only trout for Johannes, I knew, though I consoled Ida, for she was very unhappy with him.

He did not attempt to disagree with what she said; it was partly true. He only grinned, shook his head, and took a sip of grappa. Then he lit a cigarette and offered me one.

"*Truchas son la locura de el,*" Ida cried.

"What?" Johannes said finally. "You did not enjoy the trip?"

"I understand what you're saying," she said. "It's a joke."

I woke the next morning with a severe headache. It was our last day of fishing before we returned to Austria and I somewhat regretted having drunk so much, yet I enjoyed the occasional hangover as it made me deeply introspective.

While Ida stayed behind to walk on the beach, Johannes led me through more devastated villages to a beautiful river called the

Krka. I wanted to fly-fish but decided to dive with Johannes instead, as the cold clear water looked inviting and Johannes encouraged it. "It is my favorite river for diving," he said.

I dressed in all the cumbersome gear, wet suit, weight belt, fins, snorkel, mask, and lowered myself into the river. My wet suit immediately filled with cold water and my body began to warm it. Johannes had already been in the river for ten minutes before I made my first pass down the long clear pool.

I saw many trout, but was distracted by the shiny brass artillery shells that littered the bottom, and an old refrigerator. I came up to Johannes under the bridge, where he was sticking his head inside the body of a sunken automobile. He led me by hand signal to the fish that he had chased there and indicated with his hands that it was a very big trout.

When I peered into the passenger-side window the biggest trout I'd ever seen, more like a salmon, was looking back at me, its eye gleaming like a quarter. There was no way to catch it so we took turns sticking our heads in the car body to look and then coming up for air.

"It's a meter long," Johannes said when we surfaced. "A soft-mouth trout too. Too bad she is so far inside, otherwise I could catch her."

This was not only a trout but a veteran of war.

We arrived in Sankt Veit, Austria, that evening at the small apartment over the bakery where their children, Benedikt and Mariela, watched television.

# The Secret Den

During the day I was the prisoner of Johannes's son, Benedikt, who was two years younger than me. He had recruited me as his opponent in playing murderous blood-spilling computer games that I thought were a waste of time. I escaped to my room, which was Mariela's room, to take a nap.

The second evening we were home, Ida had a friend to the house to do her hair, a handsome blond woman named Rika. The woman dyed the gray roots from Ida's hair and brushed it straight, complementing Ida on its fullness.

Johannes had been out all day and returned while Ida and Rika were having coffee in the kitchen. His month's worth of reddish beard had been shaved, what hair was left on his head trimmed, and he was visibly drunk. Ida yelled at him for not telling her where he was going, and for not spending time with me. I felt, I believe, as Johannes did, that more than anything I wanted time to myself.

Beyond the door to Mariela's room was another small room in Johannes's apartment that was always closed. The next morning the door to this room was ajar and, as no one was home, I saw that as my invitation to enter and snoop. As I slowly pushed open the door I saw an entire wall of stuffed trout hanging with labels beneath each, describing what they were and where they were from. There were perhaps thirty of them, taxidermied skin mounts, well executed and museum quality, collected from all the places Johannes had been. At a glance I saw fish from Morocco, Spain, Greece, Turkey, Bulgaria, Hungary, and Yugoslavia.

There was also a whole wall of books, a library of information about trout in Italian, Russian, German, and French. On a desk in the corner was an inkwell and a pen and some paper. He was writing a letter to a biologist at the university in Barcelona, Spain, and beside the letter were large detailed topographic maps of Turkey that I had never seen.

During our trip I had thought Johannes was following a simple road map, while the reality was that for months he had been studying detailed maps and memorizing the locations he wanted to visit. When I saw the maps I immediately realized why he had not brought them; they were Turkish military maps, and if we had been caught with them they may have raised suspicion with the soldiers about our purpose. I had underestimated Johannes's preparedness, the care he had taken, and I had greater respect for him.

I don't know how long I was in Johannes's secret trout den, but while I was there he returned from work and found me. He did not seem surprised, he just stood in the doorway, holding a leather pouch like a purse. He let me speak first.

"These fish are amazing," I said, pointing at the trout on the wall.

"Yes," he replied, "but I wished I had the trout from the Tigris." He paused. "Do you want to go with me to the bank?" he asked.

He had collected the monies from the three family-run Schöffmann bakeries in town and was on his way to deposit them.

I joined him and on our way back we stopped at a bar for a beer, then another bar for another beer. When we returned to his apartment Ida was cooking dinner, a goulash with meat, local forest mushrooms (*Eierschwamm*), and fresh *Knödel.* Johannes sat me down with another beer in the living room and then went to fetch a few things to show me.

He brought me his first passport. I saw from the date beside his portrait as a young man that it was issued when he was twenty. Its pages showed, like a map of his itinerary, his trips to Colombia, where he had apprenticed as a master pastry chef in Bogotá, and other travels, mostly to Africa. There were frequent visits made to

Croatia and Slovenia, each border crossing marked by a colorful stamp. The young man in the photo had more hair and his face was leaner and longer, but it was unmistakably the face of the man I knew with the mustache, the sly smirk, and the monomania for trout.

Next Johannes brought out maps of Italy and the neighboring islands of Sardinia and Corsica and showed me his plan for an expedition there in search of trout. He wanted to make it in September and asked if I would join him. Among other things it happened to be on my parallel; of course I would.

Early the next morning I found outside my room that Ida had washed and folded all my dirty clothes from the trip. She had begun calling me *hijo,* her son, and I felt equally close to her. They had invited me into their home as a kind of permanent houseguest, and because they were so pleasant I did not feel like I was imposing.

When I walked into the living room I saw that Ida had placed a framed photograph of a beautiful, young, and slender woman on the television set. On closer view I saw that it was herself as a girl of maybe seventeen. I knew the photo had not been there the day before and suspected she had put it there for me to see. Later that day reading Dante I encountered the following passage.

> Within that land there was a mountain blessed
> With leaves and waters, and they called it Ida;
> But it is withered now like some old thing.
> (*Inferno,* XIV 1.97)

In the afternoon Johannes and I went to the local swimming hole, a lake where young women lay out topless on long sun-exposed docks. Ida did not like it when Johannes went to Längsee, but he did nearly every day, to nap and swim after his long morning of work in the bakery. Johannes and I swam out nearly to the other side of the lake and then swam back and fell asleep in the sun.

On the way back to Sankt Veit, Johannes and I stopped for a beer at a bar called Sonnhof.

"*Prost,*" Johannes said, raising his glass to mine.

"*Prost,*" I repeated, drinking the cold beer.

As we drank, Johannes reminded me of the organization of illegal fishermen we had founded, the *Schwarzfischers,* at the Gostilna Žvikar in Slovenia before we'd left on our long trip. "We will have to be *Schwarzfischers* on our next trip especially," Johannes said. "You cannot just fish anywhere in Italy like we did in Turkey, there are rules, private water, and wardens. We have to be more careful, more like *Schwarzfischers.*"

Being a *Schwarzfischer* was a complex duty. A *Schwarzfischer* fished wherever he wanted, regardless of the signs posted on the tree, but carried an obligation to be purposeful while doing so. We felt that our cause, or our individual causes at least (mine being to document the trout of the world in watercolors), transcended governments, borders, regulations, and treaties; an arrogant assumption, but, we felt, a justified one. We took our cue and purpose from the last line in Tortonese's paper:

> Further researches are badly needed for a better knowledge of morphology, variation, habits and distribution of the Trouts of Europe and Asia.

## Italy—On the Road to the Home of Enrico Tortonese

One September morning, sans Ida, Johannes and I left for Italy and the Mediterranean coast for a two-week trout-hunting expedition. It was not long before we were drinking Rosso d'Astura (cheap but good wine, three dollars a bottle) and eating thin-crust mushroom pizza in the town of Civitavecchia on edge of the mild sea.

"Specimen collection differs from sportfishing," I said to him. "A sportfisherman can enjoy himself without catching anything, but a specimen collector who has caught nothing has failed. I will enjoy myself on this trip even if we see little."

At dark we caught a ferry to Golfo di Aranci, Sardinia.

The night was warm and it was refreshing and curing to smell the salt air again. Johannes had brought some biological papers concerning trout for me to look through. I read by the light of a small low-watt bulb in the cabin we had rented for the overnight trip.

After recording his scientific findings on excursions through Turkey, Enrico Tortonese returned in the autumn of 1952 to his post at the University of Torino's Zoological Institute, in Torino, Italy. There he pondered the morphological differences of the fish he had caught and observed in drainages of the Black, Caspian, and Mediterranean seas. For comparison he added information on the trout of Sardinia, the localities of which I recorded in my journal.

The language and typography of the paper appealed to me. I thought, there is something very official in science, an order rarely encountered in other aspects of our lives. It sanitized my spirit to think that the infinite biodiversity of the trout could be formalized and placed into neat species and subspecies categories. Of course it could not. Nevertheless, I was excited to see what the trout of the islands of Sardinia and Corsica looked like.

Johannes was restless, and had trouble sleeping; he tossed and grunted in his small bunk as the ferry mildly pitched. It seemed he had not yet shed the hurried rhythms of the baker—rolling, kneading, folding, cutting, baking, cooling, all for a daybreak deadline—let alone the schedule of working in the wee hours of the morning and sleeping away part of the daylight hours. But on the ride in the car that day, I had noticed signs that he had begun his metamorphosis from baker to explorer.

At daybreak we made landfall in Golfo di Aranci. At the docks we watched fishermen bring in their overnight catches. Buyers met them with wads of paper lire at a kind of clandestine auction of Neptune's bounty.

The fishermen held out trays of silver glinting sardines and innumerable other fish they'd dredged up in their nets, small red croaking ones and eel-like ones. Other fishermen shook debris from their nets as the soft rose light of dawn played through them, casting weblike shadows on the decks of their boats. Small fisher-boys with rods in their hands, their skin African black from the sun, gawked at the professionals from the docks and begged for scraps to use for bait on their lines.

When the selling had finished, the fishermen mended their nets with spools of thread. The docks smelled like fish, and cats ambled through the pilings. Dry lightning flashed over purple-colored hills to the south where Johannes and I were headed.

Johannes had showed some pause since we had hit land. He stopped at a nearby café and we had an espresso, and watched old men, each at his own table, crumpling the pages of the paper, *il messaggero.*

As we drank our coffee, Johannes explained that we were headed to the Fluminedda River, one of the streams Tortonese mentioned in his paper. Shortly we were back on the road.

Approaching the town of Òlbia the palm trees were no longer as blue as they had looked in the early-morning light. I could now see lime greens and cadmium yellows at the roots of the fronds and a light hazel brown on the trunks. Johannes said that our knowledge of Spanish would go a long way communicating with Sardinians. The country was under Spanish occupation in the sixteenth century and the language had been infected. "For instance," said Johannes. "The Italian for river, *fiume,* in Sardinia it's *riu,* more like the Spanish *rio.*" Ida could say what she wanted about Johannes being singularly minded, but the more time I spent with him the more I saw he had a penchant for languages.

When we had left Sankt Veit it had been raining and cold. Here, as the sun rose, the heat increased and our bodies and minds were tricked into thinking that summer had returned. We came into Orosei, a red-roofed village where elderly women wore black and young women did not appear to exist. We bought some sheep cheese, pecorino, two bottles of red wine, and a loaf of *ciabatta.* The town was encircled by a fortress of opuntia cacti that bore red, prickly fruit.

Just south of Orosei was the valley of the Fluminedda River, the first on our list of streams. In the dry heat of day we started up an unmarked road hoping that it might lead to a bridge where we could have a look at the river. Small dark grapes in tight clusters were hanging among yellowing leaves in the vineyards ready to be harvested.

We found the Fluminedda River and looked off three separate bridges. What we saw was a streambed of smooth stones that, although dry, gave off an aura of damp coolness. We wondered if Tortonese had led us astray until we reached the fourth bridge, where at last there was water, though it was not moving.

We started up a dirt road and pulled over when we thought we had gone far enough. We had parked beside another Land Rover, which had a sticker advertising the Italian Speleological Society (speleology is the study of caves). It seemed that other things besides native trout brought people here.

For our hike up the river, Johannes took his mask and snorkel. I grabbed my fly rod and a chunk of *ciabatta* and we set off through a forest of large, prickly leafed oaks, meandering around their roots and trunks.

We walked up the dry creek bed as if it were a path and after a half mile or so we heard running water. At first the river ran only as fast as the sweat off our faces and we wondered, could trout live there? Then we reached a gurgling run and a small waterfall. I put my hand in the water and it was warm, like bathwater. I dipped a stream thermometer in the water. It was 83°F, a temperature that ordinarily was lethal to trout, though I knew of some trout in desert streams of Nevada that had evolved to live in such conditions.

We left the streambed and took to the oak forest for a good mile, listening for the sound of cascades. A few times I thought I saw the reflection of water through the filtered light on the forest floor, but then I realized it was sweat from my brow that had rolled into my eyes and blurred my vision.

The rushing sound of water became louder as we walked back toward the stream's edge. The banks were no longer hard and dry but suddenly moist and the air was filled with the scent of lemony mint. Beyond the lush mint patch we saw a long pool reflecting the green leaves of a giant plane tree. As we looked beyond the reflection we saw a black shape dart across the pool and knew that if we had indeed seen it, the only thing it could be was a trout.

Johannes backed off from the pool and stripped to his swimsuit. He put on his mask and snorkel, grabbed his net, and eased into the water. He swam to the opposite bank and lodged himself between two large rocks, where he groped for several minutes under the roots of a plane tree that helped form an undercut bank. Finally he lifted up his net and in the bottom of it a dark, almost black creature wriggled.

"Bravo," I said, clapping. Johannes walked to the bank where I stood and emptied the contents of his green net onto the lush mint. It was a stunning fish unlike any trout we'd seen, with tiny irregular speckles like cracked pepper on a yellow field. Why such a small fish would fill us with jubilation was beyond any evolutionary justification that I could surmise. I thought of the speleological fanatics floundering with torches in some area cave and wondered if a rare bat would bring them such pleasure. Such was the nature of a *loucura* no matter what the subject.

In other pools upstream we saw more trout, though I was not able to catch any with my fly rod, and then the water disappeared, like a snake retiring to its hole and we were standing on a dry riverbed again.

At Johannes's frantic pace we made the southeast coast of the island in time for dinner. Women and young girls were line dancing to live music at the restaurant where we ate and we could still hear the music from where we camped on the beach in Santa Lucia.

The next morning we saw flamingos wading in the marshes along the road. The air quickly became Africa hot. I was astounded by the climatic change and how different it was from my home, though it was roughly on the same parallel, 41°N. After a breakfast of fresh bread and espresso, we drove to the interior toward the mountains.

We gained elevation and passed through a region called Terrasoli. The air became cooler. Thorny scrub grew to the very tops of the mountains, thick and seemingly impenetrable. Cork trees grew on the hillsides by the road, their bark recently harvested as high as human hands with knives could reach (incidentally, when the bark of the cork tree is harvested the tree does not die; the cork grows back).

In the valleys, chestnut trees grew tall with wide trunks and were hung with clusters of fruit protected by spikey husks. We passed no one on the roads for miles. I began to realize why they called this place Terrasoli, or lonely land. The only trace of humans we saw (besides the harvested cork) were signs posted to the trunks of trees: *Divieto di Caccia,* No Hunting. The sign we wanted to see was *Divieto di Pesca,* No Fishing, because then we at least would know that fish lived there.

Searching for trout on this trip was not like it had been in Turkey, where we recruited old local men to help us. Here Johannes wanted no one to know our business. We had to rely on the scant information we had and our own wits.

We were driving down a poor road along a dry streambed when the sky suddenly turned dark and dime-sized hail began pelting the car. We stopped the car under an oak tree to wait out the storm. When the clouds passed we stepped out of the car to see the ice accumulated on the ground. We heard running water but after several minutes of searching could not find out where it came from. Had we imagined it?

"If we have not found a stream by this evening," Johannes said, "we will seek local knowledge in the bar."

In the mountain town of Aritzo we took a break from camping and got a room in a small inn. Near dark, we walked the streets of the village to the first bar we found. The men inside were huddled around a television set watching soccer. One or two of them heard Johannes and me speaking Spanish and took an interest in us.

"Where have you come from?" asked an old man holding a bottle of Moretti beer.

"I am from the States," I said in Spanish, "and he is from Austria."

"It's a long way from Aritzo to home then," he said. "We welcome you. What have you come for?" He glanced over at the television set to see if he could catch the score. Johannes hesitated to tell the old man.

"We are trout fishing," I volunteered, and then, almost at a whisper, "Do you know where we might catch some?"

"You have not picked the best time of year for trout fishing," he said, "you can see that the rivers are very low. I'm sure you've seen."

"That does not bother us," Johannes said. "The trout must still be there."

"They are there," he agreed, and blinked his eyes heavily. "I can't really tell you where, though. I fish only twice a year, and when the river's up. But only when I get the taste for them. We are so close to the sea, you know, there are enough fishes to eat."

"If you want to know about trout," another man said, "ask the bartender."

"Yes, he is the real fisherman," said the first man drinking from the bottle, and then he spoke to the bartender.

"I can draw you a map," the bartender said. "I think that would be easiest. It's near my home on the Flumendosa River."

"They are black spotted and their bodies are yellow?" Johannes asked the man.

"That's what we're looking for," I said.

"Yes, yes, they look like that," he said, drawing the map on a bar napkin and then explaining it. The bartender gave us each a beer and did not charge us. The old man had treated us to a round.

At breakfast the next morning in the inn overlooking the distant mountains, Johannes searched for the map the bartender had drawn for us on the bar napkin. To his horror he realized that he'd taken it out of his pocket that morning and blown his nose in it.

"Shit," he said, and chuckled, "it's in the trash in the room."

"I remember what it looked like," I said. "It led us right back to where we had been in the hailstorm."

"I noticed that," said Johannes, "but then he drew an arm off the dry streambed."

"We should be able to find it then," I said.

*Çildir Stream, Caspian drainage, Turkey.*

We spent the day walking beside a dry streambed. I did not mind the hike; even if we did not find fish it was a beautiful land. Autumn was near, you could smell it, dry chestnut leaves in warm, dry air. The trails we followed were made by wood pigs, which we heard snorting and occasionally saw. It was easy to mistake the sound of the wind through the trees for running water.

Suddenly we saw it, glinting like silver as it cascaded from a deeply shaded grove of oaks. The smell of the air had suddenly turned damp, of mushrooms and moss. The pools of the stream were too dark to see into and dried oak leaves choked the eddies. We did not know if we had found the place we were told to visit, but it looked as promising as any place we'd seen.

I proved with my fly rod that there were trout living there. A small fish came up to hit my dry fly. The fish was almost entirely black from living in the peat-stained water, like an eel or a night sky.

Johannes was elated. He put the trout in a plastic bag filled with water and carried it alive back to the car, where he photographed it in a long narrow glass tank. When he was done he killed it, sketched it, and removed its liver to preserve in alcohol. A sense of remorse crept over me when I thought of how freely the trout had taken my fly and how freely Johannes had taken its life. But I enjoyed holding the fish in my hand, studying it and painting a small watercolor from life as it lay on my paper casting a small violet shadow.

As we left the valley of the Flumendosa, into another I knew not the name of, I felt at peace. The roads in the mountains were winding, the cutbacks so severe they almost doubled over themselves like a snaking river that touches at its elbows and leaves oxbows.

We spent two more days in Sardinia and then, having circumnavigated the island in a clockwise fashion, we arrived at the north where the ferry was loading for Bonifacio, Corsica.

# Indulgences of a Corsican Kind

I sat at meals of wood pig, cheese, seafood, with wine of both tannic red and clearer golden, in country restaurants in the mountains and by the sea. As Johannes had described it, traveling in Sardinia and Corsica was pleasure by indulgence and not a strange pleasure through sacrifice and fear, as our trip through Turkey and the Balkans had been.

Johannes described Corsica as bountiful.

"There is more water," Johannes said on the ferry from Santa Teresa to Bonifacio. "The food is better too, because it is France. The trout are now *truite,* not *trota.*"

Riu Solenzara was the first Corsican river I saw, like a necklace of clear turquoise pools tumbling down through a rocky valley. In places the banks were open, in others there were tall pines, which appeared to be very old. The only drawback, in Johannes's eyes, to the natural beauty of Corsica was that it attracted a lot of tourists. Many of them were bathing in the river as we drove up the Solenzara Valley.

"Shit, there are many people here," Johannes said when we pulled off in a grove of ancient pines. "You'd better leave your fly rod here. You don't need your wet suit either, the water is warm."

I grabbed my mask and snorkel and walked to the river behind Johannes. We stripped to our swimsuits and slid into the pool. The water was pleasant, like tepid bathwater, and clear. I stared at the bottom and watched the light play over the river stones.

I dove down and searched for trout in cuts in a ledge on the opposite bank. I looked under rocks and slipped along the bottom pretending I was an otter. I surfaced to see if I could spot Johannes and instead I saw a young couple, completely naked, making love on the rock ledge.

I dove to the bottom again and looked under several big boulders hoping to see a trout. I scared one out of its hole and chased it to the ledge, where it hid in a small niche. Under the water you could not see the trout's colors; the fish looked subdued and the same color as the slate gray ledge. It was only when you took the fish out of the water and beheld it in the sun that you saw its brilliant colors.

I tried to find Johannes so I could show him the trout and he could try to catch it. As I swam upstream to look for him I passed two naked women swimming breaststroke across the pool. I felt somehow that I was trespassing in their territory. They didn't acknowledge me formally but they were giggling when I came close to them, maybe because I was wearing a mask and snorkel. I felt like a little fish beside them.

I don't know how much time passed, but when I found Johannes he was standing by his glass tank, already photographing a trout he had caught.

"Where have you been," he asked, "looking for nymphs? I have seen some too. We have finished work early, now we can go eat."

We ate a good meal of *cochon du bois* and set up our tent by the beach, where we slept soundly.

The next morning was Sunday. We drove up a river called the Golo to search for trout. The farther we drove the steeper the ledges on either side of the river and the narrower the road that followed it. The dawn bathed the canyon walls in a rose light as if the sun was emanating from the river. Farther upstream, the canyon leveled out and we went through a forest of tall pines, Forêt de Valdu-Niella, where we heard the distant shots of hunters. The floor of the forest was blanketed with dry needles, and paths were cut through by forest pigs.

We fished a tributary of the Golo in the dark forest that gurgled over pristine colorful river stones. The sun had not reached this spot and the air was still cool.

Johannes put on his wet suit and dove in the small pools, thrashing like an alligator as he chased after trout. He warned me that there were park wardens around and that I should not fly-fish, so I sat on a large boulder above the creek and read. After some minutes, I looked at the sun making its way higher in the sky, this brilliant orb I had seen from every place I'd ever been.

Before late morning when the sun had reached our cool quiet spot, Johannes had caught three trout. They had yellow sides and mostly red spots, like some trout we had caught in Turkey. Johannes carried them alive in a bag of water back to the car.

By noon we were heading back down the canyon to fish another spot where the Golo River had collected several tributaries and was larger, with deep emerald pools. On the way, we pulled off on the roadside by a stand of large chestnut trees to eat some bread and cheese, and to photograph the trout in Johannes's glass tank. A warm dry breeze blew through the shady stand of trees and made music in the chestnut leaves.

I was hungry and sat down to eat, cutting hunks of cheese with my pen knife.

"Remember that spot in the poplar grove where we had tea with the Turkish man?" I said to Johannes, who began setting up the materials he needed to study the trout.

"Yes." Johannes chuckled. "I remember. His name was Celal, wasn't it?"

"That's right."

"That was a nice trout you caught there in Balik Çay."

"It was," I said, reminiscing. I asked Johannes if he thought we should open a bottle of wine to celebrate the three trout we had caught that day.

"Let's save it," he said. "We have one more place to fish today." I handed him a piece of cheese to eat.

"Oh, hell," he said, "why don't we drink the wine now." We ate more cheese with the wine and finished the bottle in the car.

All along the winding mountain roads were signs of unrest from Corsican separatists. The graffiti spray-painted on barns and road signs read *libertà per Corsica.*

"We are always searching for trout in places where people want their independence," Johannes said, referring to the Kurdish people who wish to secede from Turkey. "The Basque country is another place where there are interesting native trout and separatists."

We were driving slowly through the village of Castirla when I spotted a sign on a building that read *Chocolaterie,* the *C* in the shape of a fish hook with a small fishing fly. I asked Johannes if we could stop and investigate.

I walked into the *chocolaterie* and spoke with a small woman who stood behind a glass case of chocolates. I asked in French the best I could why there was a fishing fly on the sign. She said it was because her husband was a *pêcheur de la mouche,* a fly fisherman. I told her that I was too.

"He's fishing down in the river now," she said, "nearer to Francardo. Maybe you will find him."

Francardo was a typical Corsican mountain village, gray stone buildings right up to the street, built so tightly it looked as though no mortar had been used. Small clusters of cosmos and michaelmas bloomed in cracks in the slim sidewalk where there was one.

Though the air was still cool, an occasional warm breeze blew from the river valley. The rock ledges and boulders had been baking in the sun all day and now radiated the heat like a second source, warming the air.

Before fishing, we drank a cold Coca-Cola at a small café.

"It's a good day for swimming," Johannes said.

We parked our car just upstream from the village and walked down a steep and rocky hill to the river. The water in the river was tepid, and when I put on my mask and dived in I saw many trout. They held stationary in the current like fruit in a bowl of Jell-O, as if the movement of the water did not affect them.

I must have made a discreet entry into the water because many of the fish continued to feed with me there, darting to catch small insect larvae floating by. I was mesmerized watching them as I tried to swim in place. I could see the trout were different from ones Johannes had caught in the upper tributary that morning. Their sides were peppered with fine black and red spots and their mouths looked like they were shaped differently.

When I had swum enough I put on my shirt and pants and began walking upstream, hoping to find the fly fisherman whose wife ran the chocolate shop. I had not yet seen a fly fisherman in Italy or southern France.

After about fifteen minutes of walking I saw an old man sitting under a fig tree, the broad green leaves like three-fingered mittens concealing his face. When I walked closer I saw that he was lighting his pipe and that a fly rod was leaning up against the tree.

"*Bonjour, monsieur,*" I said.

"*'Jour,*" he said.

"Have you caught anything, *vous avez eu une touche?*"

"*Non,*" he answered. He looked to be older than his wife, if he was the man I was looking for. His pipe smoke was carried to my nose by a warm breeze. It smelled like burning chocolate.

"I am a fisherman, from America," I said. "My friend and I are looking for native trout, *truite sauvage.* Do you know if the trout here are native?"

"Ah," he said, "the trout of Corsica have lived through a lot. They have introduced other French trout to our island." He puffed on his pipe. "The Nazis threw explosives in our streams to kill the trout to eat. I was a member of the *Résistance.* Do you know this, do you understand me? *The Résistance!"*

*"Je pense."*

"I know the trout of this island, what they used to look like. Most of these streams have introduced fish. In another valley there was a very peculiar fish, wild and native, with big red spots. I can show you this fish. I have one in my freezer that is fifteen years old. The trout in that stream are now gone."

"You are from the chocolate shop, yes?"

"Yes, I make the chocolates. If you want, you can meet me at the shop tomorrow morning. I will show you this fish."

"I would like to see the trout. I will be there with my friend in the morning."

I left the man to his fishing and walked back down to the pool where Johannes had been diving.

I told Johannes about the old man and the peculiar native trout he spoke of that is now extinct that he kept in his freezer.

"This guy is old," I said. "I think he was in the war."

"Which war?"

"World War Two."

"Did you tell him we were fishing?" asked Johannes.

"No."

"Good, you never know, he could be a warden."

"I doubt it," I said.

We drove over the pass to the coast to spend the night and had dinner in the town of Porto. We started at the bar with a beer and then drank several glasses of pastis on empty stomachs. I spoke about my excitement to see the frozen trout.

"So, it's just a frozen fish," Johannes said, "I prefer them live."

"But it's a native fish and they're no longer around. The man said they were unique."

"How does he know, he's just a *chocolatier.*"

"And you are?"

Johannes smiled. "Well, we'll see, maybe he knows what he is talking about."

For dinner we ate *salade niçoise* and thin linguini with mussels in a white wine and garlic sauce. It was nice to be up in the mountains and by the coast in the same day, neither one very far from the other here. Corsica was not a very big island.

After dinner we had another drink.

"We have covered a lot of territory, you and I," Johannes said. "We have drunk pastis in five countries by five different names." He was talking about the clear anis-flavored drink we had that turned cloudy when you added water. Johannes listed them. "Raki, *aslan sütü,* ouzo, *Anis del Mono,* Pernod, pastis, Ricard."

We went without speaking for some time, watching what went on around us. The sea air had a bit of September melancholy in it.

The next morning was our last in Corsica. We had to make the town of Bastia by evening to catch the ferry to Livorno, Italy. But first we had an appointment with the old fly fisherman at the *chocolaterie* in Castirla.

"The trout of Corsica have lived through a lot," the old man repeated when I entered the shop with Johannes, though I think he was talking more about himself. He was wearing an apron and wiped his hands on it.

"I suppose you've come to see the trout," he said. "Well, I'm sad to say I looked for it all yesterday afternoon and into the night and could not find it. I took everything out of my two freezers and put it back in three times over." The old man took his pipe, which had been sitting in an ashtray, and lit it. "When I asked my wife if she had seen my trout she said she threw it out with some other old fish five years ago. I'm sorry, monsieur, it was a large trout with big red spots. I'll tell you what they looked like, the spots were like ripe raspberries. But, can I give you some complimentary chocolates to take with you?"

I took some chocolates that he had put in a small wax-coated paper bag.

"Thank you, monsieur," I said to the *chocolatier.* Johannes and I headed for the door.

"Typical of a wife to throw out the old man's trout," Johannes said as we were leaving.

"Hold on," I said and went back inside.

"Can I ask you one more thing?" I said to the man. "Do you have any locally made flies for trout fishing? I would like to bring one or two home as a souvenir."

"Why, yes," he said, taking his pipe out of his mouth to speak. "I have the best Corsican flies, made with feathers from my friend's prize gamecocks. Let me go in the back room and get one or two for you. I always keep my fly-fishing rod nearby."

The old man gave me two beautifully tied dry flies that were blue gray in color, the stiff hackles of the feathers bristling from the hook like an insect's legs. I thanked him and left the chocolate shop.

Johannes made a point of reminding me that I had cost us some time, that if we had not stopped at the chocolate shop then we could have fished another stream on the way to Bastia. Now we had no time. We caught the overnight ferry to Livorno, Italy, at twilight.

We made landfall at dawn and saw a red sun rising over a field of dried corn as we drove through the city of Pisa. From the hillside above town we could see the famous leaning tower glowing red with reflected light, like a piece of penne pasta cloaked in marinara sauce. We were headed up into the Apennine Mountains in north central Italy to fish one last stream before returning to Sankt Veit.

Johannes spoke briefly about not wanting to return to work. Then he began talking about how it was mushroom season and where we were fishing might be a good place to find *Steinpiltz,* or stone mushrooms. His tone was peaceful as we spoke about other places where we wished to travel.

"If I can get the time off, I would like to go with you on your latitude trip to Central Asia. There are many interesting trout there that I have not seen. This will take a lot of planning, but I have good information on the trout, and a few contacts. It is best to know people, if you can arrange it.

"I hope you will stay in Sankt Veit as long as you want. We have rooms on the third floor that are usually occupied by young bakers who are training here, but they are open now. No one is using them. You can have an apartment and we can plan our travels."

"I'll take it," I said, referring to the third-floor room, "but I'll have to repay you somehow."

"You can buy me a beer," Johannes said.

So I became a temporary resident of Sankt Veit an der Glan.

## SANKT VEIT: BARS 80, POPULATION 10,000

Sankt Veit had a special tax for buying flowers to adorn the central square, its own torte (called the *St. Veitertorte*), and a city wall built in the thirteenth century. That summer, for the second time in fifteen years, Sankt Veit had been voted the most beautiful town in Austria. I found it charming and clean, a model place to live, but many of its residents found it isolated and boring.

I had settled comfortably into my third-floor room and had made it a bit like home. I hung up clippings and photos and some of my paintings. I spread out sprigs of dried flowers and leaves that

I had collected. Before me on the desk I put a single fig leaf, on the hard surface of which I wrote a poem, the first poem I had written in a long while.

Johannes took me on several weekend excursions to Slovenia, where the mountain rivers were getting colder and the marble trout were preparing to spawn. We visited a place he knew, on a tributary of the Soča, where very large marble trout were digging redds in the dime-sized gravel (a redd is a kind of nest where the trout lays its eggs and then buries them). The spot was at an overlook by a ledge. On the top of the ledge there was some ice I had not seen, and I slipped and fell part of the way down. Thankfully, I did not hit my head on a rock, but I banged up my left knee and for the whole day I was limping, though I thought nothing of it.

I fished several times for a large salmonlike fish called a *Huchen* that were purported to live in the nearby Drau, a large tributary of the Danube (the fish is sometimes called the Danube salmon). The sun no longer gave any heat; it was cold, and occasionally it snowed. I did catch one about five pounds and a brown trout of equal size, casting and retrieving large shiny spoons. These catches encouraged me to continue trying for one of the very big fish, like the one I saw hanging on the wall of a Sankt Veit pub.

Johannes spent his spare time writing of his scientific findings during our travels that summer for the journal *Österreichs Fischerei.* He also had cleared three months for travel the following summer, and we were both working to secure visas and permits for travel. The countries we were considering, Armenia, Azerbaijan, Iran, Uzbekistan, Kyrgyzstan, Kazakhstan, and Mongolia, were not all easy to get into. Several travel advisories dissuaded Americans and Europeans from travel in remote regions of Central Asia.

I spent most of my time with Johannes and Ida and their children, Mariela and Benedikt, but my companions were not all members of the Schöffmann family. I had made friends with some other

Sankt Veiters. One was Klauss, a tall blond student of abstract philosophy.

One night out at the bar where the big *Huchen* hung on the wall, I got very drunk and was introduced by Klauss to the bartendress. She was a young brown-haired girl with brown eyes, a round face, and slim body. She gave me a complimentary drink. Johannes showed up and had a drink with Klauss and me. The girl and Johannes had looked at each other, talked in German, and laughed. I was not too drunk to notice that I was being set up.

I stayed until the early morning when the bar had closed. I watched the girl clean the bar top, stack the clean glasses, load the washer with dirty ones, and then she followed me to my room and spent the night. I was afraid that people would know I had a girl in my room, but then I didn't care. She was passionate and eager and her wrists smelled like alcohol. The cool Austrian, almost autumn night was beckoning through the window. I was experiencing a form of companionship and intimacy I had not felt since my stay with Yannid in Rouen.

I shared with the girl, whose name was Alexandra, my master plans for travel that winter and the following summer, my maps, my drawings of fish, my journals, the conglomeration of clippings and photos and pages written on with notes I hoped to distill into a book. She took an interest, and for the next two weeks we spent a good deal of time together. All the observations and thoughts I had been holding inside I told to her. She spoke English well. I took her fishing on the Drau and explained the angler's hope of catching a big fish.

She was leaving soon to study in Vienna but continued spending time with me up until the day she left. When she did I wished her a good journey.

"It's not so far," she said, referring to Vienna. "You will have to come visit and we will go skiing."

"I will," I said.

I was slightly solemn when she had gone and Ida teased me.

"Ah," Ida said. "My son is in love." I suppose I had spent enough time above the bakery to be adopted.

October left and November came. The cornstalks in the valley at the foot of the Alps had turned golden and been cut. The life in the trees had been drawn into their roots. The first flakes of snow fell, collecting on the insignia for the Schöffmann bakery on the wall of the building outside my window, a large pretzel, and on the cobble street. I already was dreaming of warmer days and trout fishing. I was happy, arranging my fishing equipment on my bed, taking inventory of my flies.

But one morning I woke in my bed and as I was sitting up to look out the window onto the rooftops of town, I noticed that my left knee had fluid in it. I got up and went about my day, read, showered, and tried to ignore it. An hour later the amount of fluid in my knee had doubled and I had trouble bending it. I walked down the three flights of stairs, and by the time I got to the bakery where Ida worked I was having trouble walking. Ida asked me why I was limping.

"I don't know," I said, "it just happened."

"Too much dancing," she laughed.

At night it throbbed like a bad tooth and I experienced a rush of nightmarish thoughts as I lay awake. The next morning there was so much fluid in my knee that I couldn't see my kneecap and the skin around it was stretched to the tautness of a drum.

After two weeks of hoping the fluid would go away on its own, the swelling only became worse, so I decided to seek medical attention. I thought of the fall I had taken off the ledge in Slovenia when we had gone to see the marble trout spawning in the Soča. Maybe I had strained it, had walked too much and too rigorously, had not rested it.

The doctor I visited did not know what was wrong either. He said to wait a few days and if my situation did not improve he would remove the fluid with a needle.

Those few days passed with no improvement, so when I returned, the doctor inserted a needle below my kneecap and extracted nearly a liter of yellow fluid. I left his office with a bandage around my knee and by afternoon the joint had filled up with fluid again. I returned the next day and the doctor did not venture to guess what was wrong but sent me to a rheumatologist.

As I walked down the streets of Sankt Veit trying to conceal my limp, people asked me what was wrong. "I hurt my knee when I fell off a ledge," I said. "It's injured." The word *injured* for me carried in its connotation a hope for recovery. "Or maybe," I said, "I picked up a strange parasite in Turkey that affects your joints. It's a tough bug, but I'm going to beat it."

I could get around with some pain, but I began to worry—what would happen if my other knee went out too?

Already I had trouble doing basic things, kneeling down to sit on the toilet or getting in the backseat of a car, because I could not bend my knee. These were new challenges I faced. I became irritable because I did not want to be delayed by having to struggle with so many petty things that should come easy. I refused to think of all the things in life I had taken for granted, the most immediate one being the ability to walk.

The rheumatology specialist I visited was named Peter, someone I recognized from some bars in Sankt Veit. I looked to be the only person under seventy in his waiting room.

When the doctor saw me he lay me on an examination table and aspirated my knee with a long needle, following that with a cortisone shot in the same place.

"I know you from the bar, don't I," the doctor said. "You are Hannes's friend."

Afterward he sat me down and asked a series of questions. "Do you have a family history of arthritis, are you allergic to any medicines, do you have psoriasis?"

"Psoriasis?" I said, "like the dry-skin condition?" I told the doctor yes, I did have a mild case of psoriasis.

"Ahah!" Peter said, with an enthusiasm I would not have expected.

He looked at me over his reading glasses and squinted, speaking with a heavy Carinthian accent. "I think I know what you have. This," he said, putting his hand on my swollen joint, "is an imperfect knee." He leaned forward slightly. "I think you have a rare form of arthritis related to psoriasis called psoriatic arthritis. We don't know much about it, how you get it, what the relationship is between the psoriasis and the arthritis."

"I don't know anyone in my family who has had this problem," I said.

"There are some severe forms and cases," the doctor continued. "But I am hopeful yours is monoarticular, that is, it will attack only this one joint."

Peter handed me a color pamphlet about the disease. It contained pictures of symptoms, like pitted fingernails and red sores all over the body. It spoke of extreme cases, people whose every joint—wrists, knees, elbows, and knuckles—were swollen. "Sausage fingers" was one term that described swelling in the hands. As a painter this scared me most of all. In truth, the doctor had made me petrified with fear and dread.

"I never heard of arthritis in young men," I said.

"There are things we can do," the doctor said. "I'm putting you on a strong dose of an American anti-inflammatory drug. Don't walk on it if you can help it. Take it easy."

"Doctor," I said, "I live in a three-story walk-up. I hike, I fish, I like to strain my body. How bad is this going to get? Will I be confined to a wheelchair?"

"I'm not sure."

"You're not sure? Will I be able to hike and run and walk along rivers?"

"You should," he said, and stopped making notes on my chart to look at me. "Just not maybe as fast as you used to."

Everyone around me, including Johannes, asked what was wrong, why my knee was swollen, why I was limping so severely. At the bar, people asked Johannes, "What's wrong with James?"

"He hurt his knee," Johannes said.

When they asked to see it and I lifted up my pant leg I saw disgust in their faces, especially in their eyebrows. I had never experienced being the object of such extreme pity. I hated it. I had no solace except the hope of recovery and the thoughts of all the places I wanted to visit when the weather warmed.

I began to envy those who could walk without limping, even the people I viewed as the ugliest, most miserable souls, and tried to hide my own limp but could not. I knew people could see that I was limping because I could notice it myself in my reflection in the windows of shops in town. I had an intense and increasing fear that I chose not to face, that which told me I might not get better.

In this state I would not be able to do many of the things that were so important to me. I could not help thinking that even the laziest stream would suddenly be an obstacle to me, and that my life was over as I had known it. I began to despair when I thought that I might even get worse, would not be able to paint because my fingers would be swollen, or hike mountains because I could not bend my knee, or walk down a steep embankment to a river, or even leave my bed.

Ida was openly sympathetic, called me her son, and told me I would get better. Johannes could not understand why I was not right, but never did he suggest that my condition would jeopardize our plans for the next summer. "You must get better," was the only thing he said to me, and he said it only once.

Johannes and I studied maps, sometimes daily, in his secret

*Man with char in the snow and* kuma zsasa, *Hokkaido, Japan.*

PROSEK '99

front den, and he gave me scientific papers to read (though I could not read all of them, as some were in German and Russian). The images of the trout in these scientific works, the sometimes primitive drawings in Russian journals published in the nineteenth century, gave my recovery a purpose; I wanted to see those fish in the flesh and paint them on my own.

In the wake of a cortisone shot in my knee the swelling would go down for a couple of days. Then it would fill up with fluid again. Toward Christmas and the New Year my knee was in good enough shape that I could dance a bit, but my mind was so obsessed with the idea of complete recovery that I had little fun. I did not go out much, and essentially for the winter hibernated with my thoughts and memories. That is, until I saw Alex, the girl bartender in Sankt Veit who had returned from Vienna for Christmas break.

"Get over it," she said to me in the bakery one day, more or less, "don't languish like a pussy, everyone's got problems." And she walked out the door.

Later she came to my room and told me about her semester at school.

"I'm sorry you can't come skiing with me," she said. "You *will* get over it."

She stayed with me that night and her tenderness gave me hope. That was just about all I could ask for.

I did get over it. Though I was not back to normal by the time I packed my bags for a trip to Japan in late April, my knee was good enough that I could walk around without much pain or a noticeable limp. I feared doing something that would set it off again and was overly conscious not to bang or twist it in any awkward way.

After months of limited mobility I had been given freedom to walk again. My recovery coincided with spring and that made it even more magnificent. It was then I realized that the adversity I'd faced that winter was something I had secretly wished for my whole life, a fault or imperfection that might help to push and challenge me and wipe complacence away forever. I hoped even more deeply that my problem would not return and continued taking the drugs the doctor had prescribed for me.

## HOKKAIDO, JAPAN, 41°N—A GIRL WHOSE NAME MEANS LITTLE RIVER

My next journey on the latitude was with my friend Dawn Ogawa, who was tall and thin with straight black hair. In Japanese her last name meant little river. When she graduated from Yale (where we had met and dated) she returned to her home in Hawaii (where she grew up with her Japanese father and American mother), and shortly thereafter left for Japan on a Fulbright Fellowship to do cancer research in a hospital in Kansai.

When Dawn invited me to visit her in Japan, she of course knew that my idea of a trip would involve trout. "That's okay," she said through e-mail, "it will give me an opportunity to see a countryside I have not yet seen."

We chose May for our travels because Dawn would be on holiday, and I researched where we might find good fishing.

An indispensible resource in the planning for our trip was a man named Katsuhiko Yoshiyasu, an ear, nose, and throat doctor from Kyoto. As with Johannes Schöffmann, he was introduced to me as a trout specialist by Dr. Robert Behnke. Behnke had received a copy of Dr. Yoshiyasu's beautiful new book on native Japanese trout and encouraged me to contact him. I wrote him and began a correspondence, sharing books and photos concerning our mutual passion.

In my last letter to Dr. Yoshiyasu, sent from Austria in the midst of my miserable state, I expressed an interest in fishing Hokkaido, Japan's northern island, explaining to him that Hokkaido was on the 41st parallel, which I was writing a book about. As the doctor did not speak or write English very well I told him that further arrangements would be made through my Japanese friend in Kansai.

Dawn wrote Dr. Yoshiyasu in March and he replied to her with a twelve-page handwritten letter. He included maps (favorite spots marked with a yellow highlighter), appropriate flies, meticulous diagrams showing how to fish for the native trout, and pictures of the fish with the names spelled in phonetic Japanese on the back. Also included were his regrets that he could not join us. "Hokkaido is beautiful," he wrote, "though it's a little cold in May."

I arrived in Kansai the week before the boys' day holiday, meeting Dawn in the airport and taking the train with her to the suburb called Nishinomiya, where she was living. The symbol of boys' day is the fish, and from every high pole, rooftop, and car antenna colorful fish flags were flying against a blue sky.

As Dawn showed me around Nishinomiya, I admired the colorful fish flags and noted that fish were just as at home in air as they were in water. Tubular in shape, the flags filled like sails and swayed in a slight breeze or darted in a stiff wind.

I also observed the many rock gardens in people's yards. "They're called *Kare Sansui,* dry mountain stream," Dawn said. "They're supposed to express the spirit of Zen through only rocks and sand. The sand around the rocks is usually raked into a design to create the effect of movement like current in a stream." The fish flags and the rock gardens resembled each other, I thought, in that they both represented the beauty of water, through suggestion, without water.

Within the walls of the backyard gardens were also small bonsai in pots and some larger trees trimmed, wired, and trained to grow in a manner that suited the gardener's taste. Dawn told me that the controlled trimming and shaping by the Japanese evolved out of a desire to make their landscapes look like those in Chinese brush paintings and create the illusion of distance in a small space.

Dawn's host family, the Tanakas, lived a short distance from the train station. At the door to the beautiful home we took off our shoes and entered. Dawn led me to a small room with rice paper walls and a small lamp where I could put my things. Her host mother, Mrs. Tanaka, was busy preparing dinner—homemade pumpkin soup, stir-fried spinach, squid sashimi, chicken, and fried rice with pork—and did not come out immediately to greet us.

When we ate, Mrs. Tanaka complimented me on my deftness with chopsticks. "Dawn taught me," I said and smiled. There was a clean cool smell in the room like there is in the air filtered by a deep forest.

Mrs. Tanaka was confined to a wheelchair. She had lupus (a disease that affects the skin and joints) and for the past ten years had little to no use of her legs or feet. For five years she crawled to get around the house because the floor was made of woven reeds, tatami, which would be destroyed by a wheelchair. Then, the greater Kansai area had one of its worst earthquakes on record. In Nishinomiya, the earthquake came at 5:30 A.M., when people were still in their homes. Many people died, but Mrs. Tanaka was an ironic beneficiary. The damage created by the earthquake gave her and her husband, an elementary school music teacher, insurance money with which to remodel their home, now friendly to Mrs. Tanaka's needs.

Mrs. Tanaka's passion was cooking, and during my stay there I ate many home-prepared Japanese meals. She spoke little English, but was able to communicate to me when slicing raw squid one day that she was holding her favorite sashimi knife.

After dinner, Mrs. Tanaka helped set up the futon I would sleep on. My room was the only one in the house that still had the traditional tatami mat floor. She could not ride her wheelchair in so she propped herself out of it and crawled about the room setting out this sheet and that pillow, carefully building up my nest. I sat on the floor with her and showed her photos of some Japanese char (a fish similar to trout with light spots as opposed to dark spots), and streams they came from. She nodded at their delicate beauty and their colors, perhaps wondering to herself how she might prepare them. She mouthed the names of the trout, almost at a whisper, written by the doctor on the back of each photo in Japanese:

*yamame*
*miyabeiwana*
*amemasu*
*oshorokomo*

Dawn and I left Nishinomiya on the morning of the first of May for Sapporo, the capital city of the island of Hokkaido, north of Japan's main island (called Honshu). My limp had all but disappeared, and though I could still feel fluid in my knee I could walk and carry our camping equipment without a problem.

The airport was jammed with people because it was the first day of Japan's spring holiday, "golden week."

As Dr. Yoshiyasu had predicted, Hokkaido was cold in May. An overcast sky and cool breeze chilled Dawn and me though we wore several layers of clothes. We walked around the city of Sapporo browsing in bookstores, where I bought several beautiful books on fishing in Japan. It was apparent from photos that the average trout caught by a Japanese angler was small, probably six inches. Dunking flies in small fern-covered pools, however, seemed an appropriate extension of an already well-established aesthetic, for enjoyment by the Japanese of small things in small places was well known and documented. For me fishing small brooks for small trout was the purest and most enjoyable form of angling, and therefore I felt at home in Japan.

There were three main freshwater trouts caught by Japanese anglers—the landlocked cherry salmon, *yamame;* the Dolly Varden trout, *oshorokomo;* and the white-spotted char, *iwana.* The words *yamame* and *iwana* were usually spelled out in hiragana (the phonetic alphabet) but there were, in addition, special kanji, or characters, for them. The kanji for *iwana* combined characters for mountain, stone, and fish, presumably because they lived in rocky, mountain streams, while the kanji for *yamame* taken separately were mountain, woman, and fish. Why woman was part of the kanji for *yamame* was something I was interested in pondering further. In Latin languages, trout was always feminine—*la trotta, la trucha, la truta, la truite* (and also German, *die Forelle*). I later learned that the Japanese character for woman was present in many other Japanese words, like "noisy" and "inexpensive."

Dawn and I planned to sample the nightlife of Sapporo before we headed for the countryside. She contacted a fellow Fulbright scholar living in the city to ask how we might pass some time. Mark, a tall, dark-haired American about our age from Utah, said he would be happy to show us around. He arrived at our hotel with his Japanese girlfriend, a beautiful young woman who modeled for fashion magazines and karaoke videos.

We drove in her BMW to a lookout above the sprawling city.

"Before the mid-nineteenth century," Mark explained, "there were no Japanese living on this island. It was occupied by a native people called Ainu. Sapporo was founded in the late 1800s and many of the municipal buildings were built in the colonial New England style because a man from Massachusetts named William Clark was employed to help organize the government and plan the city."

The four of us ate dinner at a *tabehondai,* an all-you-can-eat restaurant, and then we left for a karaoke bar downtown, where we sang and danced until morning.

After several hours of sleep and slightly hungover, Dawn and I began our drive out of the city of Sapporo. All signs of urban life gave way to wide farms of newly plowed dark soil.

We stopped at a small teahouse and had some green tea, which smoothed the kinks in my system from the night out. As we moved toward the island's interior and into higher elevations, snow still covered the fringes of fields and the earth lay dormant. Soon we had left the flat arable land and were traveling through forests above rock ledges. At intervals along the road, white-winged birds with black masks appeared and disappeared. They seemed to be leading us up and into the snowy hillsides to the tree line and over the pass where groves of white birch became gnarled and stunted.

The lower-elevation streams ran high and discolored from silt suspended in the melted snow but the high tributaries ran clear and looked like perfect trout streams. Seeing them run from springs and melting snow was like watching my own blood start to move again. The river was my best tonic.

We followed directions the doctor had given us to our destination, a national park on the Shiretoko Peninsula, Japan's easternmost point.

Near sunset we stopped in the coastal town of Abashiri by a river lined with fishing boats. We asked a fisherman if he'd had any luck. "I have not been out this day," he said, "but if I had I would have been fishing for *tara* [cod]. The season is still young. Heavy ice floes from the Amur River [Russia] prevent us from leaving the harbor until March and even now it is tricky in the sea."

A big red sun was setting behind him over jagged peaks to the west.

* * *

Early the next morning, Dawn and I started out for the mountain road that wound up over the center of the Shiretoko Peninsula. The sun was bright on the ridge in the white landscape where we stopped to have lunch in a rustic wooden lodge, an udon noodle soup with shrimp tempura. Both the sea of Okhotsk and the Pacific Ocean could be seen from this point at the top of the pass. The jagged coastline resembled Maine or, as Dawn said, her native Hawaii.

After lunch we descended the southern slope of the peninsula. As we approached the rocky coast we saw fishing villages built on small spits of land at the bases of cliffs. There was a light surf that tossed the fishing boats at their moorings. Salt spray blown by the wind wetted piles of nets on shore.

The craggy cliffs were draped with metal curtains and cables that prevented snow and rock from falling on the fishing towns that held tenaciously to the thin coastline.

In one of the villages, called Rausu, Dawn and I checked in at a *minshuku,* or bed and breakfast. The price of the room included a *washoku,* which we had after our fishing that evening; it was a big dinner of crab legs, fish-head soup, and several small dishes, mushrooms, scallops, octopus, and local woodland root vegetables, served in small red-lacquered bowls.

Dawn and I took off our shoes and entered the *minshuku,* escorted by a small lady to our room. We put our things in a corner, unrolled each of our futons on the tatami floor, and took sheets and blankets out from the sliding *washi* paper walls that concealed a closet cavity. I tested my hard pillow, stuffed with dried beans. I checked some of our maps and the doctor's photos. Dawn lit a small wood fire in the stove to stave off the damp chill air of the coast.

I held a map to my face and peeked up over the top to look at Dawn and remembered that she was beautiful and I had once loved her. She proved a good travel partner, a *Schwarzfischer* even.

"I'm having a good time," I told her. "I like Japan, I feel at home here." I put down the map and pointed to a stream. "I think we'll fish here," I said. "According to the doctor, this stream has all three species of the trout I want to catch."

We told our host that we would be back for dinner by seven and made our way to the Chmishibetsu River, as it was called.

We crossed several small rivers entering the sea on the way to the Chmishibetsu. None of them was very wide or very long—they started at the ridge on the peninsula and the peninsula was very narrow.

We found the river and parked on the bank near its confluence with the sea. The estuary was littered with bleached-white shells, fish skulls and spines. The river swelled with snowmelt and tumbled and rolled with noise and force over a series of falls. I grabbed my fishing rod, and Dawn and I headed upstream through banks of deep snow.

We walked through shoulder-high thickets of a green bamboo-like plant called *kuma zsasa,* bear grass. New green shoots poked up from the soggy ground and melting snow. Along the way up the river I peered into small clear rivulets that entered into the main roily river.

We came to a quiet dark pool that looked to be of melted snow, shaded by tall stands of bear grass. I looked into the pool between the rotting leaves and sunken logs and saw small dark fishes swimming around. I was inwardly elated. The fish in this pool were trout, I was certain.

I pointed the fish out to Dawn. They appeared to be slightly agitated and swarming like flies.

"There are many," I said, "all about the length of my hand."

"I see them," she said, "they are very dark."

"Their backs are," I said, "but their bellies will be red, like a fire engine."

I crawled to the edge of the pool on my hands and one knee, and sat to look in my box for an appropriate fly. I tied on a small dry fly, one that had been given to me by the Corsican fisherman the previous September, and cast into the pool from a very awkward position. The slate-colored fly floated and, as there was no current, did not move until it disappeared in the mouth of one of the little dark fishes.

I pulled on my line and the fish came tumbling out of the water and into my lap.

"Its belly is so orange," Dawn said. "How did you know it would be?"

"It is a char, and when they are in dark water like this they seem to have accentuated colors. Most char have red bellies at spawning. And these light spots on the sides are also unique to char."

The belly was not so much red as it was orange, the way Dawn had described it, like squash blossoms. My relationship with Dawn had begun in a painting class at Yale; we were therefore intimate with each other's style and taste. Dawn admired the fish, as I knew and hoped she would, with a painter's love of pigments and materials. "It's like a flame," she said.

"This is a subspecies of *Salvelinus malma,* the Dolly Varden char." I said more specifically, "*Krascheninnikovi,* indigenous only to the Shiretoko Peninsula where we stand." I looked at it lying in the snow. "But I suppose the name doesn't matter, does it."

Dawn and I were careful and quiet by the pool. I handed the rod to Dawn and showed her how to cast it. She quickly caught one herself. We thought about keeping some to eat, but after admiring them briefly in the snow, we returned them to the pool. Then we left the small pool that we had momentarily disturbed.

Upstream, the banks of the river became steeper and the snow deeper. A light drizzle began to fall. Because I wore waders I was able to walk upstream in the shallows of the river's edge and fish quiet eddies in the roots of trees. Dawn could not follow without getting her shoes wet so she stayed behind and took out a small sketchbook to draw the river.

I walked farther, as I have had the urge to do since childhood, feeling the way I used to, as if every bend in the river ahead of me were a new world. I walked over the uneven ground, and not without some trouble. I hesitated to leap off a rock, and my knee felt strained, but the long miserable winter in Sankt Veit was being separated from me by new experiences and time.

My face was wet with the falling drizzle and I felt as if I was in a very remote place. I thought I was alone, and then saw fresh footprints in the snow.

I followed them, heading away from the river through the bear grass. Coming down a slope of crumbling rock, trying too hard to be careful, I slipped and fell. On my way to the ground I crushed my rod, but when I inspected it, it did not appear to be broken. When I looked up I saw whom I had been following. It was a small man with a peaked hat and overcoat of straw, and over his shoulder, slung on a sprig of bear grass, was a large char.

The man was making progress through the woods. I wondered where he was going, as he was heading up into the mountain and not toward the road.

## Chasing the *Sakura*

At breakfast the next morning at the *minshuku,* Dawn and I talked to an older couple about our outing the day before, how we had not believed the amount of snow still mounded by the river in May. They lamented that spring had been very late that year—the vacationers were disappointed with the cool wet weather—but they had just come up from southern Hokkaido and reported that the *sakura,* cherry trees, had begun to blossom there.

The blossoming of the cherry trees in Japan is a celebrated event, recorded widely in words and paintings. Dawn and I both wished to see this.

"It is the ultimate sign of spring," she said.

As we drove away from the peninsula, to the south where we hoped to find warmer weather, steady rain began to fall. Some large waves crashed over the coastal road on which we drove. We continued to head south all day, arriving that evening in the town of Shizunai. Though we had brought camping gear it was still too cold. We found a cozy and reasonably priced *minshuku* at which to spend the night.

After a warm shower, we took a walk through town. Along the streets were trees with pink blossoms.

"*Sakura,*" Dawn said. "Why didn't we see them when we first came into town?"

I looked at the blossoms, reached up, and touched one.

"It's plastic," I said. We found out at dinner that the fake blossoms had been hung by the townspeople as a kind of apology to those who had come on their holidays for the cherry blossom festival.

In the afternoon of the next day, driving along the sea, the overcast sky broke open and some sun beat onto a raging surf, showing that the water was not always gray and smoky, but transparent and blue green. We were determined to drive south as far as we could in order to see the cherries in bloom.

We covered a lot of ground that day and by evening were near the southern tip of Hokkaido. According to Dr. Yoshiyasu's map, the area, besides perhaps having blossoming trees, also had a good concentration of banner trout streams.

We checked into a *minshuku,* a beautiful wooden lodge that resembled a ski chalet above a pine-covered hill. The inn was not far from the village of Nisseko and sat at the foot of a large snow-covered mountain called Yotei-zao.

Dawn and I took our shoes off at the door and when we walked in we were greeted by the smiling faces of other people boarding there. The walls of the *minshuku* were hung with gorgeous photographs of local wildflowers. The owner of the lodge was a wildflower specialist and all the boarders in the lodge were there for the spring flower watching. We had walked into a kind of secret convention of others who shared a common *loucura.* They were the *Schwarzfischers* of the stamen and petal, though I felt that they were mysteriously kin.

"What flowers?" asked Dawn in Japanese, when they spoke of flower watching. They all laughed in unison.

I had not seen any flowers either, so I said, "What flowers?" in English. They all laughed at both of us. Spring came swiftly, they told us, and you had to be there waiting before it came in order to see it, or you might miss it. *It* was when *It* happened. *It* was the magnificent time of transition in the natural world, when energy shot back into tired flu-bedraggled limbs. *It* was a harmonic orgasm in all things, something to be awed by, and *It* did not last very long.

Drinking rice wine after dinner the second night of our stay, a couple told us they had seen the cherries blossoming that day in Hakadote, a town on Hokkaido's southernmost tip (on the 41st parallel). The next day we planned to go fishing there.

Early the next morning the clouds had returned, and when we woke and rolled out of our futons it was raining. At first there were no breaks in the gray sky, but as we drove south the rain diminished and patches of pavement on the road became dry. We stopped for lunch in the town of Onumae beside a pair of lakes and there we saw the cherries in bloom.

Across the small cove was a tree half in bloom, its horizontal limbs painted with intermittent pink highlights. The distant hills were splashed with small strokes of pink too. I could not ever remember having seen so much pink in a wooded landscape.

The day unfolded as an answer to secret wishes. The display of soft-colored petals against the gray sky grew more and more spectacular the closer we came to Hakadote at the southern tip of the island.

We took the highway along the ocean and then turned inland up the Shiriuchi River. As if all at once before our eyes, the world had turned green; the forests on the foothills leading up to the snowy crags were soft and light like moss, dotted with the pink petals of the cherry blossoms. The land was rejoicing.

I had chosen a tributary of the Shiriuchi called Detofutamata to try fishing and we followed on the map as best we could to find it.

We passed a man and his wife who were photographing flowers by a tiny brook and stopped to speak to them. They found it amusing that we were after trout. We drove as far as we could up the mountain road near to cliff edges and crossed over a small rushing brook on a rickety wooden bridge.

"We might try here," I said to Dawn.

I rigged up my fly rod and after some minutes of fishing below diminutive cascades in dark pools I could not see into, I had caught several *iwana,* a small silver-blue char with large white spots. I held the fish in and out of the water and photographed them.

"You look at them like that couple looked at the flowers," Dawn said.

That evening at the lodge Dawn and I told stories of the fish we caught and the *sakura.* After dinner we joined several guests at a nearby *onsen,* or hot spring bath. Stripped down to our bare selves sitting in a warm natural pool, I closed my eyes to the cool dark evening. Imprinted like stars on the insides of my eyelids were pink *sakura* blossoms and the white spots of the small native trout.

The next morning the owner of the inn at the foot of the mountain took Dawn and me to see a patch of yellow bell-shaped flowers that had just opened up. He did not photograph them because he already had taken thousands of frames of this single place. Apparently it was a rare flower and very special to him. The petals were delicate like tissue paper and the whole patch fluttered in the slightest breeze.

Days later, Dawn and I returned to the main island of Japan. We visited the temples at Kyoto and walked through the perfect gardens of stone and maple where moss is worshiped and weeded by tiny ladies hunched over like snails. I bought some art materials at a district in Kyoto, Gosho Shimogamo, known for its beautiful handmade papers, sumi ink brushes, and pens. I wanted to teach myself and partake in an ancient Japanese tradition of making ink prints of fish on rice paper, called *gyotaku.*

Dawn and I ate lunch in one of the temple gardens, and she presented me with a gift. She had made me a *hanko,* or a stamp made of wood, with my name, Prosek, carved in phonetic Japanese. It came in a small box with a stamp pad of vermilion ink.

"I enjoyed our trip together," she said, "you made me see something I would never have seen on my own. Now when you make *gyotaku* you can put your name on it, in Japanese."

*Johannes de la Mancha, Spain.*

# Part III

# Spain Again, and Portugal

After the long winter, my first trip with Johannes took place in a country where we both spoke the language tolerably well. On my return to Sankt Veit, over a glass of beer in the Sonnhof bar, Johannes and I finalized plans for a trip to Spain and Portugal.

Collecting native trout in Spain would be especially difficult because the last pure populations were scarce, closed for fishing, and highly protected, and we could face severe fines if we were caught by a warden. If we had more time we could have probably arranged, for scientific purposes, to fish legally, but we had only three weeks to sample some thirty or forty small rivers in all corners of Spain and Portugal.

Our source of information for the locations of the last native trout was a graduate thesis on the trout of Spain, published by a student at the University of Barcelona in 1994. Johannes had been in correspondence with the student, whom he called simply Garcia, who now taught at the same university. Garcia gave Johannes detailed information in exchange for information on Turkish trout.

We left Sankt Veit on a rainy afternoon. Three days of driving, and four languages later—German, Italian, Piemonte, and French (each with its own word for trout, respectively, *Forelle, trota, truta,* and *truite*)—we approached the Pyrenees Mountains on the border with Spain (and three more languages, Catalan, Basque, and Spanish—*truita, amuarrain,* and *trucha*).

I had not thought of our itinerary as flexible and then I remembered how dependent our success was on the weather. As we approached the town of Perpignan we slid beneath a ceiling of clouds the color of dark grapes. We had hoped to coordinate all our trips to encounter the least rain, the warmest weather, and the least number of tourists. June had seemed to be the right time, but the distant sky, streaked with rain like falling razor blades, made us feel that we had not been as clever in our planning as we had thought.

"Rain makes trout hunting difficult," Johannes said, "it turns the water off-color. But we have a choice," he added, pulling over and taking out the map. "When we reach the city of Perpignan we can go south, clockwise around Spain, or counterclockwise as we had planned."

"I think we should go south first," I said.

"Why?" he asked. "Well, anyway, I agree, then we'll end up in the Pyrenees, where it is colder and wetter, at the end of the trip, and by then maybe it will be dry and warm."

We chose the land of oleander, olives, and oranges.

As we neared Valencia, I saw an orange tree and then groves of them, and felt the warm sun on my skin. I was happy to be a passenger, watching the stillness of a world going by, lulled into a meditative state by the repetition of passing trees and fence posts.

It was not long before we had left the highway and were heading up into the mountains, a pattern that was familiar to me now after several trips with Johannes. Native trout now lived in remote places, where few people did.

The Rio Linares was the first stream we fished, flowing east out of the Sierra de Gúdar toward the Mediterranean Sea. It ran through an arid hill country below clusters of terra-cotta-tile-roofed homes built on small peaks and ridgelines. We sampled the stream at several points along a ten-mile stretch. Because of the rough terrain, there was no single road that followed the course of the river, so we took no-name dirt roads that might lead us to the water.

We were encouraged when we finally saw a sign posted on a tree that read *Vedado de Pescar,* Forbidden to Fish. The sign, to a *Schwarzfischer,* was an open invitation.

I packed my seven-piece fly rod in my backpack and we hiked downstream on a tractor path. When I arrived at a good-looking pool, concealed by lots of brush, I took the rod out of the backpack, put the pieces together, strung it up, tied on a fly, and fished. I had become very efficient at assembling and disassembling the rod quickly. I caught two small trout in Rio Linares, handing them to Johannes, who held them in a plastic bag filled with water as he kept a lookout for wardens. I continued to fish.

"We have enough trout to photograph," Johannes whispered through the bushes, "put your rod away."

"I just want to fish a minute more," I said. "You take me all this way and put me on a gorgeous trout stream and expect me to fish only for five minutes?"

"There will be other streams," Johannes said.

"Not like this one," I said.

"Better," Johannes said, "don't be greedy, there may be wardens."

The trout I caught on diminutive dry flies were yellowish green with fine red and black spots. They also had three or four dark verticle bands along the sides, a characteristic I was finding was typical of many trout from streams flowing to the Mediterranean. I had seen these peculiar markings on trout from eastern France, northern Greece, Croatia, and Corsica. I had told Johannes that the French called such trout *la truite zébrée,* zebra trout. We discussed the trout over dinner that night at a small restaurant in a small mountain village run out of someone's home.

"Not all Mediterranean trout have it," Johannes said, referring to the zebralike bands, "but most do. I have never seen these *zébrures* on trout from Atlantic drainage streams or on those from the Black or Caspian seas. It is unique.

"The Iberian Peninsula is interesting for trouts, in the same way that Turkey is," Johannes continued. "In Turkey you have rivers going to four major bodies of water, the Caspian, the Black, and the

Mediterranean seas and the Indian Ocean. In each drainage, trout have been isolated for thousands of years, and in each they are different. In Spain and Portugal you have rivers going to the Atlantic and the Mediterranean. I will be curious to see how the fish differ. The trout of southern Spain interest me most. I believe they will be like trout I caught in Morocco."

A fascinating puzzle was unfolding before my eyes. In almost every stream we fished, the trout were slightly different. Because of the nature of the way brown trout evolved—oceanic ancestors isolated by retreating glacial melt, and confined to specific rivers for thousands of years—their diversity is astounding.

The next day Johannes and I headed farther south on winding roads to the basin of the Tajo River, the longest river in Iberia (its headwaters were in eastern Spain and it flowed to the Atlantic in Portugal). The location we wished to fish was mentioned in Garcia's paper as being near the village of Peralejos de las Truchas, about two hundred kilometers east of Madrid, in the province of Guadalajara. The name of the stream with pure native trout was Rio de la Hoz Seca, or river of the dry gorge. In his thesis, Garcia included the name of the local warden who was in charge of protecting the stream.

When we had driven a good way into the mountains and the roads had become narrower and more difficult, we began to wonder if we had taken a wrong turn. The road at times seemed too small to lead even to a farmhouse, let alone a village. But eventually we arrived in Peralejos de las Truchas.

It was a village built of stone, not far from the Tajo River. We drove through the narrow cobble streets and found that the only roads beyond the village were four-wheel tracks. It made sense that the predominant vehicles in town were old box-shaped Land Rover Defenders.

The main section of the Tajo near town was absolutely clear and you could see many trout holding in the current and rising to flies. According to Garcia's paper, though, nonnative brown trout had been introduced and hybridized with the natives. The original genetic strain in the Tajo was no longer pure. The tributary we wanted to fish, however, Rio de la Hoz Seca, was isolated from the main stem by a waterfall, and Garcia wrote that this section had never received introductions.

Our map showed that Rio de la Hoz Seca was about thirteen kilometers outside of town, but as far as we could tell, no roads went there. Left with no other source of information, we were forced to ask locals which track might lead to it. The problem with asking for information about Hoz Seca, though, was that it might raise suspicion with the locals.

"We don't want to draw attention to ourselves," Johannes said. "Somebody might inform the local warden."

"You really think that the local people know there's a special protected trout there?"

"You never know," Johannes said. "You have to assume they do." He laughed. "It's the only town I've ever seen that has 'trout' in its name."

We parked in the small central plaza in Peralejos de las *Truchas* and walked around until we found a bar. In these small villages the bar seemed to be the source of all local knowledge, at least that's how Johannes perceived it.

Inside the bar were three old men, shouting in a singsongy dialect. They weren't shouting out of malice but just seemed to be hard of hearing. We sat down beside them and felt the bar shake as they rattled their glasses on it.

*"Dos cervezas,"* Johannes said to the barmaid.

When she brought our beers, Johannes asked if there was a place to stay nearby.

*"Sí, hostal de Tajo,"* she said, "it's very nice." She was wiping down the bar with a rag and Johannes got her attention again.

"*¿Sabe usted,*" Johannes asked, "*si hay carretera a la hoz seca?*"

"Sure," she said, "just past the church, take your first left. Then at an old abandoned farmhouse take a right. At the bottom of a steep hill the road will split again and you go left. The road will cross the river in the *hoz seca.*"

I noticed behind the bar that there were some fishing lures and flies for sale.

"Is this town a fishing destination?" I asked the woman.

"Oh yes, that is our main business, serving fishermen. But people also come here to hike and hunt."

"Can you buy a fishing license in town?" I asked.

"The closest place to get a license is Molina, thirty-eight kilometers away. It takes a week to process the license forms, though, because they have to mail them to Guadalajara, the capital of Guadalajara. Or you could go to Guadalajara yourself, which is a three-hour drive from here. They don't make it easy. And that license is valid only in the province of Guadalajara. If you wished to fish in Andalusia, Asturias, Galicia, or País Vasco, you would need other licenses."

"*Dos cervezas mas,*" Johannes said, finishing his beer. The woman brought them and an old man took her place. He put out complimentary tapas in front of Johannes and me, plates of fried pork rinds, olives, and fresh anchovies. Johannes and I were eating the tapas and drinking our beers when a man walked in and sat beside us.

He noted us as foreigners, ordered a beer, and introduced himself. "I am from the nearby town of Checa," he said. "My name is Jorge. What brings you to Peralejos?"

"Just sight-seeing," Johannes said.

"Oh."

"I have been curious, though," I said to the man. "What does the *Peralejos* mean in the name of this town, Peralejos de las Truchas?"

"I think *peralejos* is an Arabic word, but I can't remember the story. Do you know, Pedro?" he said to the bartender.

"No," the bartender said.

"I've probably drunk too much absinthe. It erases the mind."

"Do they have any absinthe here?" I asked. "I've always wanted to try it, it's illegal in America."

"Of course, yes, yes, they do, don't we, Pedro," he said to the old man.

The old man shook his head. "No."

"Sure," the man from Checa insisted, "you must. Señora," he said to the barmaid, now sitting in a corner, knitting, *"tienes absenta?"*

*"Sí,"* she said, and walked to the bar and pulled out an old wine bottle with a liquid in it and no label.

*"Tres,"* Jorge said, and the barmaid poured out three glasses of green alcohol.

Another man walked into the bar and saw our drinks being poured.

"I get drunk just looking at that stuff," he said. He was wearing a fly-fishing vest over a camouflage shirt. He sat down at the bar and ordered a beer. He had left the door open and I noticed that it was getting dark outside. A cool breeze blew across the bar.

"Did you catch anything?" I asked the fisherman.

"Yes, I had very good fishing just near town."

"What kind of flies do you use?" I said.

"Oh, soft hackles," he said.

"You tie your own?"

"No, I buy them from a friend in Peralejos, a man who owns another bar, called the Jube. He sells beautiful flies there, but his bar is closed now because his wife is nine months pregnant and they are staying with her sister in Molina, closer to the hospital."

The bartender put out more plates of food, roasted red peppers, smoked fish, and whole cloves of cooked garlic still in the husks. I added some water to my glass of absinthe and the clear green fluid turned milky. It tasted sweet and anise flavored, like raki or ouzo.

"I will show you on a good map where to go," said the man from Checa to me. "If you are interested in fishing."

The absinthe was strong. I had read that if you drank enough it had a hallucinogenic effect, due to its main ingredient, wormwood. It was not long before I was having some trouble speaking.

"We are interested in trout fishing," I said.

"No we are not," said Johannes, who was a little drunk too.

"I am," I said. "I am interested in trout."

"Well," said the fisherman, "trout are very beautiful creatures."

"Especially these," said the man from Checa. "We have very beautiful trout in Peralejos."

This statement left us all staring into the mirror across the bar from us.

We spent the night at the hostal de Tajo and the next morning made the journey to the Rio de la Hoz Seca. The barmaid had given us good directions and the trip was the perfect length and degree of difficulty, just challenging enough to make us feel that we were going somewhere. The rewards were remarkable. The river was one to rival in beauty any I had seen. It ran clear, though in its depths it was emerald colored, a spring-fed creek lush with green mosses and weeds. The greenery by the water was in stark contrast to the ochres and earth tones of the bare ledge and earth erupting on either side and forming the formidable canyon from which the river got its name. Most important, there were many trout.

We were very stealthy, because the stream was closed to fishing, and as I worked the banks, deep and undercut, with my fly rod, I thought, I am poaching in Paradise.

According to Garcia's thesis, the fish were what he called an ancient strain of Atlantic drainage brown trout. When I caught some and saw them, I began to formulate my own opinions.

Johannes did not dive in this stream but stood behind me eagerly, waiting to see what the fish that I might pull out of the water looked like. The trout were not easy to catch, but they were so numerous and large that it was hard to miss if you covered enough water. I caught one of about seventeen inches that fought doggedly and nearly jumped onto the bank in an effort to shake the hook.

"They look like Mediterranean brown trout," I said to Johannes when we beheld them in our hands. They were a brilliant yellow color like the sun, and we both noticed at the same time, amidst the fish's red and black dots, the verticle bands characteristic of Mediterranean brown trout.

"Yes," he said, "they have the verticle bands, the *zébrures,* but the Tajo does not flow to the Mediterranean, it flows to the Atlantic."

If you looked on a map, you would see that although the Tajo flowed west, all the way through Spain and Portugal to the Atlantic, its headwaters were very close to the Mediterranean. "It is possible," Johannes said, "that the ancient ancestor of the Mediterranean trout crossed over the divide into the drainage of the Atlantic. Or another possibility is that the more ancient Atlantic ancestor was closely related to the Mediterranean trout to begin with, that they split off from each other a long time ago and this is a relict population."

We were so excited to see this fish that was completely new to us. Johannes and I shared a genuine *loucura* for scrutinizing the biodiversity of trout, the spotting patterns and colorations of what we felt was the most beautiful of all fish.

To celebrate how successful we had been, and the startling realization that we had seen a trout from one drainage that should look like one from another, we ate some cheese and bread and a good sausage and opened a bottle of wine. When we had finished photographing the trout alive in Johannes's glass tank, we released the evidence of our *Schwarzfisching* into the river and took a nap in the tall grass laced with intermittent patches of brilliant red poppies.

We made our way south the next day, toward Andalusia, through the region of La Mancha. It was a vast, open country of golden wheat and the occasional tall dark cypress. I watched the broad yellow fields go by and the occasional big black billboard in the shape of a bull, advertising Osborne whiskey. The dry heat blowing through the olive groves lulled me to sleep.

I woke from my nap sometime later, lost in time and space, in a state of temporary amnesia. I looked at Johannes, driving with the determination of a machine down the highway and I thought, who is this mustachioed person I am spending so much time with? What is my association with him? I breathed in the dusty wheat-scented air and laughed out loud at what I began to see as our mad pursuit.

"Johannes de la Mancha," I said out loud.

By evening that day we were in the mountains again, near upper tributaries of the Guadalquivir River, drinking beer in a rustic bar in a pine forest. Heads of native deer and forest pigs hung on the walls, giving the place an enchanted hunting-lodge feel. We ordered two beers and toasted, *"Salud y pesetas!"*

My life had become a mélange of trout streams and barflies.

Beside us in the dark space, sitting at the bar and resting his powerful arm on a large oak plank that formed the bar itself, was a warden wearing a green uniform with the embroidered insignia *Junta de Andalusia.* He had biceps larger than my calves, and you could barely see his bull-like eyes through the black stubble on his face, prickly like cactus spines. Johannes nudged me and whispered, "That's good for us that he's in the bar and not on the stream."

We kept an eye on the warden through the mirror in the bar. The warden and the bartender were watching the bullfights on television. When he turned to ask the bartender to fill his glass, the warden took a long look at Johannes and me. Did he know we were *Schwarzfischers?*

Having successfully sampled a prohibited stream the next morning (a tributary of the Guadalquivir River) and caught indigenous trout, also with the stripes or *zébrure* markings, we headed to the next group of native trout streams on Johannes's list, tributaries of the Guadalquivir that flowed out of the Sierras near Granada.

I had been to Granada more than a year before when my latitude travels had just begun. At that point I had been alone and eager to procure a license to fish a stream near town for introduced trout. Now I had been indoctrinated into the cult of the wild and pure trout (and became an autonomous native trout snob).

We spent the night in the center of Granada at the hostal Perla. As I walked around town I remembered my visits to the palace of the Alhambra, watching the goldfish in reflection pools, walking through orange groves dreaming about princesses. I now felt more aware than I had been, that my senses were heightened. Why had I not seen the snowcapped mountains, clearly visible as you walked down the carril de Picón?

That night in Granada we had a bad meal for too much money and danced with tourists at a flamenco festival. Johannes expressed his disgust for cities.

We fished two tributaries of the Guadalquivir the next day not twenty-five kilometers from the city center. In the first, Rio Dilar, I caught a very strange-looking trout of a cool turquoise-blue color with very fine black and red spots. Johannes said the small and numerous spots on the sides reminded him of trout he had caught in the Atlas Mountains of Morocco.

"It makes sense that they should be similar," Johannes said, "because Morocco is geographically very close to us here in southern Spain. The fish likely shared a common ancestor. But you notice, they have no *zébrure.*"

I had caught the fish with live mayfly nymphs from the stream bottom that I had strung on a fine wire hook. The trout to me was astounding in its uniqueness. I held in my hand another piece of the evolutionary puzzle. By fishing and photographing what we caught we were collecting data, though the living data were returned to the stream unharmed. For some reason Johannes was not collecting tissue samples on this trip, perhaps because Garcia had already done genetic work on many of them.

We put the trout in the tank and observed it.

"You just know when you see a native trout. There is a look to the indigenous fish that you don't see in introduced ones."

"Yes," Johannes said. "I think when you have seen enough trouts, thousands, and looked closely at them, you know whether or not they are hybridized."

"The pure trout just looks correct."

"They are also different," Johannes added. "From every stream the native brown trout are slightly different."

We camped that night on the coast in the town of Huelva and the next day made our way west toward Portugal through a sparsely populated region of eastern Spain called the Estremadura. In this region many of the bulls headed for the corridas were born and raised. We passed them along the road, brown-black *toros* with broad muscular necks grazing under cork trees.

As we headed north through the desolate country toward the middle of Portugal, we saw the tents and carts of *gitanos.* They were not unlike Gypsies we had seen in Turkey, with large wheeled carts pulled by mules and tall conical canvas tents. The air was warm and soporific. Abandoned buildings beside the road were host to nesting storks.

We had fished the upper Tajo River days before, where it was a small trout stream. As we neared the border with Portugal we crossed it again, but here it was broad and huge. The Portuguese called it the Tejo River, and we did too when we crossed the border. We entered Portugal at Segura, a town, Johannes told me, known for its cherries. Ripe cherries were being sold by the basketful by stout ladies standing on the sides of the narrow and winding mountain roads.

The two rivers we were heading for, the Zêzere and Mondego, were small tributaries of the Tejo purported to have native trout. Johannes had garnered the information about native trout in Portugal from a paper published by several scientists at the University of Porto. The rivers were in a national park called Serra da Estrela (star mountains) with peaks pushing two thousand meters.

There were no highways across Portugal, just miles and miles of winding mountain roads. As we neared our destination, the mountains grew taller and more jagged and the roads more difficult. We came into the national park and the valley of the Zêzere River. In the village of Manteigas we stopped to ask for information on how to get to the stream. We bought a good map of the park and followed it to the upper Zêzere. This brought us to an ancient-looking place, with small villages that reminded me of Turkey with no sign of electric lines and smoke rising from the chimneys of small stone homes.

Large rounded boulders, as if strewn by a giant, dotted the hillsides, and a brisk cool wind blew down the steep canyon through which the trout stream flowed.

"Every place we visit is more beautiful than the next," I said to Johannes as he drove. He grunted in approval. I don't know if he saw the shepherd watching his flock from the top of a large boulder, or was charged like I was to see people living in this old subsistence way. He did, I know, see the water, but I don't know if he acknowledged its pristine beauty as it appeared and disappeared among the massive beige-gray rocks.

Wheat fields were tossed in the wind, farmers walked holding their hats to keep them from blowing off, everything was covered in lichen. I could not help believing that native trout lived in some of the most remote and beautiful places left in the Northern Hemisphere (trout are not native to the Southern Hemisphere, though they have been introduced and now thrive in beautiful places there as well).

Though the stream was remote, it was open and visible and we did not want to risk using a fishing rod. Johannes volunteered to dive and we found a spot hidden among the boulders where he could do this. He dove without a wet suit because he said it would take too long to put it on.

"Ah, the water is so cold," he cried in a half whisper, stepping out. He had been in it for about eight minutes and it was only about fifty degrees. But in that time he had managed to catch two small trout. The fish had characteristics of Atlantic brown trout, white halos around the black and red spots, and white rims on the ventral and anal fins. But they also had the characteristic we had identified as Mediterranean, three dark verticle bands on the sides.

Over dinner that night in a restaurant near Manteigas, we enjoyed sharing our theories on why the trout looked as they did.

We fished other streams in Portugal, the Avé, the Estorãos, and tributaries of the Lima. The deep green forests they flowed through reminded me of those behind my home in Connecticut. The farther north we drove along the Atlantic coast, the fewer the trout with vestiges of Mediterranean trout characteristics. These trout had large black spots with cream-colored halos and no trace of the *zébré* bands on the sides.

When we neared the northern town of Chavez (on the 41st parallel) in a region called Trás-os-Montes (between the mountains), I shared with Johannes a bit of my family history.

"Luiz de Oliveira," I began, "my father's maternal grandfather, was born in Chavez. He moved to Brazil in 1910 with his wife, Alicia. In Santos they gave birth to my grandmother, Amelia de Oliveira, who became Amelia Prosek when she met and married my Czech grandfather in São Paulo."

"So your father was born in Brazil," Johannes said.

*Drying whitefish, Lake Issyk Kul, Kyrgyzstan.*

"Yes, in Santos. My father always told stories about his Portuguese grandmother in Brazil. How she lived in a house with a dirt floor and let him drink coffee with lots of sugar."

For little money, Johannes and I ate like kings: home-cooked meals in small mountain villages, with delectable dried hams and cheeses, drinking refreshing *vinho verde* or newly pressed wine. Across the border back in Spain we had wonderful Galician fish soups with toad-fish, scallops, crab, and calamari, and more dried hams. When we didn't want to spend money for dinner we went to a bar and filled up on the tapas of olives, anchovies, and dried ham that came gratis with our drinks.

As we drove along the coast, I looked at the Atlantic Ocean with fresh thoughts and feelings. After spending so much time on gentle mountain streams, the beach with crashing waves seemed more formidable than before. This was the home of the ancestral trout from which all the ones we had studied evolved.

After two days in Galicia we began to head east again, toward Asturias, where there were many beautiful salmon rivers. The pace of river hopping and sampling Johannes kept was a bit frantic, though we found many healthy and well-protected streams full of native trout. We had burned through regions, drainages, languages, mountain ranges, climates, and currencies. The winding mountain roads taken from site to site had dizzied me into a timeless warp. But still, I believed that we remained true to the purpose of the expedition: to document the diversity of the trout of the Iberian Peninsula.

There was something beyond the so-called purpose, though, and that was my love of catching fish. Besides a fascination for the fish itself, I enjoyed the stalk, the capture, and the entire predatory act. I was satisfying an urge that was thousands of years old. In most cases I did not kill my prey; the catching was enough.

I liked to sit on the moss-covered banks and think about catching a trout. The anticipation was a great deal of the excitement for me. Johannes and I saw so many beautiful rivers on our way through the Basque country, beyond Bilbao and toward the French border—the Trubia, the Pisuerga, the Irati (where Hemingway trout-fished in his Pamplona days), the Nive. I wished to return to them and spend more time in each place.

For now, though, I was fixed to Johannes's agenda. I did, however, persuade him to make one cultural detour, to visit a building in Bilbao, the newly built Guggenheim Museum of Art on the Nervion River. I knew that the architect's design had been inspired by what he has described as an "obsession with fish," and therefore I thought that we should see it. The influence of the fish's form was apparent on first sight. The titanium plates of the building's exterior produced a wavy silver armor, like scales over a fish's body. As you looked at it, the building mesmerized, like the effect of staring into a pool of water. Johannes found little interest in it and drove by.

"We have no time to stop. It's not on our agenda," he said.

"Whether you like it or not," I said to Johannes, "the man who designed it shares a passion with us. He likes fish."

"Eventually, whenever I'd draw something," the architect of the building, Frank Gehry, once said, "and couldn't finish the design, I'd draw a fish as a notation . . . that I want this to be better than just a dumb building. I want it to be more beautiful. Sometimes I think fishes are all there are in the world."

We crossed the Pyrenees into France at St-Jean-Pied-de-Port, and spent two days looking for trout in brooks that percolated from thick mosses and the roots of giant fine-leafed beech trees. We had found a nice campground with good facilities on the river Nive, called, appropriately, Camping de la Truite.

As we drove through villages in the beech forests near the headwaters of the Irati River, I noticed aspects of the homes that mimicked fish, like rounded slate roof shingles assembled like giant carp scales.

Johannes displayed a boyish and scientific curiosity as he dove in the rivers, a sweet expression of the *loucura.* Our last afternoon in France, which was relatively warm, we swam in the Nive behind the campground. Johannes caught a large sucker fish with his bare hands and laughed gleefully as it flopped against his chest and slipped back into the river. As far as I could see, this was not part of the agenda.

At dinner in St-Jean-Pied-de-Port, we discussed how different our next and quite distant destination on the latitude would be. We sat outdoors and were waited on by men in white-and-black uniforms. While we waited for our roasted duck, Johannes surreptitiously threw pieces of bread from our basket to the large trout below.

# 41°N, Eastward—Yerevan, Armenia

We did not have an easy time getting to Armenia, a small country on Turkey's eastern border. Our flight from Austria to the capital, Yerevan, should have taken five hours, but instead took three days. We arrived in this desert city exhausted, nearly delerious, on a blazing hot afternoon.

At the airport we were met by two people who would accompany us overland in search of trout; a lady translator named Nuné, and our driver, Marat. I had made the arrangements by e-mail and phone and was happy to find that in person they were both amiable people. The first place we visited in the ancient city of Yerevan (founded in 782 B.C.) was the central market.

"We must get some supplies," Johannes told them.

Beneath a high-ceilinged warehouse-type building we walked alongside piles of sour cheese, dried sausage, butchered beef, whole chickens, and live carp in water-filled oil barrels. I watched the fish for some time. My exhaustion momentarily left me.

"They are from the Arpa River," Nuné told me. She was quick to answer my curiosities even before I raised them. "I like to see the fish swimming too," she said.

While Johannes was buying fruits and several bottles of vodka and beer, I rummaged through a large freezer and discovered two frozen trout. I brought them to a spot on the floor where light spilled in from a hole in the ceiling and photographed them.

"Don't waste your time with those," Johannes said, "we'll see plenty more at the lake."

"Can't I decide what I want and don't want to photograph," I snapped. We were both a little irritable from the trip.

* * *

We next had a cup of coffee at a café in the main square of Yerevan, amidst the municipal buildings that were noticeably in disrepair. Nuńé pointed out that you could see Mount Ararat from where we sat. It was barely noticeable in the haze, but once you saw it, you could not miss its peak covered in snow. There was a quiet, depressed feeling in the dry air, which approached 106°F.

"We were on the other side of that mountain one year ago," Johannes said, "in Turkey."

"I was thinking the same thing," I said.

"You travel a lot?" Nuné asked.

"He does more than me," I said.

Johannes lit a cigarette and offered one to Nuné, which she smoked. He offered one to Marat as well, but he refused.

"It should be several degrees cooler by the lake," Nuné said, taking a drag.

We sat under a red umbrella that advertised Coca-Cola. I stared, through the distortion caused by the heat, at our Russian four-wheel-drive compact car parked on the street. It would be our means of travel for the next three weeks.

"It is called Niva," Marat said, looking at the car.

"Yes," Johannes said, "they marketed those in Argentina, but in Spanish *ni-va* means *doesn't go.*" Johannes spoke in English. His English had greatly improved since I had known him.

Yerevan was the first city of a former Soviet republic I had visited. It made a strong impression. Every dark-haired, thick-eyebrowed woman and man walked with their heads hung low and many of them carried something in a plastic bag. There was a beauty in the concrete-and-steel ugliness of the town, in the inconveniences I imagined existed, but it seemed a little drab and sad (like a glimpse of Dickensian poverty). Vendors sold sugar melons and bottled mineral water. Old Russian sedans—Volgas, Ladas, and Jigulis—in white, yellow, and baby blue moved up and down the uneven streets, their tires bald and patched, their wheels missing hubcaps. It was my first taste of Russian dysfunctionalism, viewed before the silhouette of Mount Ararat, from which, according to the Bible, post-flood life began.

The lake Nuné had referred to at the café was Lake Sevan, our destination, which looked on the map to be roughly one-tenth the surface area of the country. "Though Armenia is relatively small as countries go," Nuné explained, "Sevan is still quite large. When you stare across it, it's like you're standing at the edge of the sea." She paused. "But I wouldn't know exactly, because I've never been to the sea."

Lake Sevan, the outflow of which runs to the Caspian Sea, once had four separate races of brown trout that occupied different niches in the ecosystem and spawned in the rivers entering the lake every month of the year except June. "I'm confident we will be able to find trout," Nuné had e-mailed me.

The only written documentation Johannes and I had on the trout of Armenia was a paper published in Russia in 1896, during the reign of Czar Alexander. Johannes could read the Cyrillic alphabet, so he at least could pronounce the names of the rivers where the trout spawned. Our information was unfortunately over a hundred years old and sadly out of date. Since the late nineteenth century, Lake Sevan had greatly changed. In the 1930s it was drained sixty feet for the purpose of exposing what the government thought would be arable land beneath, but the dry earth turned out to be infertile and nonarable, mostly rock and sand. Realizing its error, the government subsequently built a fifty-kilometer pipeline through the Geghama Mountain Range to bring water from the Arpa River to the lake in an effort to fill it up again.

"They won't be able to," Nuné said, addressing the subject, "but at least the level has remained stable."

The draining of the lake had caused the extinction of at least one of the native races of trout, that known as bodjak, which had spawned on the now-exposed shores. We had little information to substantiate the existence of the other three types of trout, but Nuné was optimistic that we would find them.

"The trout of Lake Sevan are somewhat legendary among Armenians," Nuné said with a prideful air. By this point we were out of the city and making our way to the lake itself. "We call it *ishkhan*, or the prince fish, and the legend is that there was once a Urartian prince [the Urartu people preceded modern Armenians] whose beloved drowned in Lake Sevan. He wanted to live with her forever so he asked a magician to turn him into a trout. That is why it is the prince fish."

One of the books I had brought with me to read was *The Hunting Sketches* by Ivan Turgenev, stories about Russian serf life in the nineteenth century. Turgenev was a landowner but wrote sympathetically about the peasants bound to his land. Looking out of the car window at the villages we passed I thought the lifestyle of the people here, subsistence farmers living in stone homes, was not much different from those I read of. The only difference was that the Armenians were not serfs, but citizens of a struggling democracy.

"People are prepared to work, but there is no work," Nuné said, sensing what I had been thinking. "I tell you, many lament the fall of Communism. They say, 'At least back then we had jobs.' Now the country is dependent on loans from the World Bank and contributions from the Armenian diaspora, mostly Armenian Americans in Los Angeles and New York. The leaders are involved in a kind of Armenian mafia. The money never gets to the people. Our biggest and newest opportunity lies in tourism, but in the West you read about how our prime minister and his cabinet members were assassinated last year, you think we live in chaos and it is not safe to come. Now there is a great opportunity to see Armenia. Westerners could not enter the country until the mid-nineties. We have much history. We were the first nation to adopt Christianity—in

A.D. 301—and we have some of the oldest and most beautiful churches in the world."

"Don't bother telling Johannes," I said, "he doesn't care about churches, he only cares about trout." I realized I had begun to sound like Ida.

"Well, trout are just another reason to be here," she said.

Eventually, in the distance, we saw Lake Sevan through a light blue haze. It lay as a kind of alpine sea, nearly six thousand feet above sea level. At first it looked no different from the pavement distorted by the heat waves, but then we smelled it, moist and not unlike the ocean, and saw clearly that it was indeed the vast inland sea Nuné had said it would be.

The lake itself was green and clear. The air was cool by the water. The shores were lined with resorts from a time when prosperous Russians took vacations there. Many of them were half finished and almost all of them were abandoned. We took the road part of the way around the lake to a peninsula, on the tip of which, Nuné said, the president of Armenia had his summer residence. As we turned onto a road down the peninsula, we encountered a man dressed in colorful costume wearing makeup on his face, like a clown. Behind him, high above the ground, his friend walked a tightrope strung between two poles. It was all very bizarre.

"Am I dreaming?" I asked Nuné.

"No," she said, "they are doing it for money." Sure enough the man walked up to the car and we gave him some dram, the Armenian currency.

Farther down the peninsula we came to a large building of overlapping concrete discs, cantilevered over a cliff, like something out of a Russian futurist's imagination. It looked precarious.

"The building is supposed to look like a fish's mouth," Nuné said.

It was one of the few resorts still operating. Marat stopped and we walked into reception to check in.

When we had settled in our rooms, each with his or her own

balcony that hung over the cliff, the four of us walked down a path to the beach for a swim. Marat and Nuné sat on the beach in the sun cutting up a watermelon. Johannes and I jumped in the water.

I lay floating on my back with my face to the warm sun and Johannes, wearing his mask, dove to the bottom to look around. He swam back to shore with a large crayfish and I swam back too to look at it.

We watched Johannes photograph the crayfish from every possible angle.

"He's a little crazy," I said to Nuné.

"I admire the depth of his curiosity," she said. "It makes life interesting."

When Johannes had finished with the crayfish, he let it go.

"Can I show you where we would like to go?" he said to Nuné. Johannes took out maps to show to her and Marat, unfolding them on the beach.

"Where did you get such beautiful maps?" Nuné asked.

"They are old Russian military maps," Hannes said.

Nuné squinted in the sun. "They are really something," she said. "It all is."

"So," Johannes said out loud, looking across the open lake. "Do you think there are trout?" He pointed to the streams that flowed into the lake where he wished to fish.

"I've arranged to meet with fishermen in the villages around the lake," Nuné said. "We will know tomorrow whether they have been catching trout in their nets lately, or if the trout are extinct."

We sat in the sun and ate slices of the bright pink-fleshed watermelon. The sweet smell was hypnotic. I put on my diving mask and swam in the warm clear water, watching the light play in all imaginable geometric patterns over the sandy bottom.

That evening before dinner, Johannes, Nuné, and I walked through a grassy meadow on the peninsula to a ninth-century stone church.

The sun was setting over the lake and a breeze was blowing from the hills behind us.

We ate dinner at the hotel, on a terrace overlooking the lake. A young girl brought out a plate of oval flat bread, called *matnakash,* and fried *siga,* or whitefish. The next course was yogurt soup and sour cheese. When we had finished the bread the girl brought more, and a plate of peppers, eggplant, and tomatoes stuffed with beef and onion. We each drank several bottles of Armenian beer, Kilikia, and at the end of the meal a glass of Armenian cognac.

I slept very well my first night in Armenia.

## PEPAN THE FISHMONGER

The next morning, having all slept well through the cool night, we visited a small fishing village called Tsovagyugh, on the northernmost shore of the lake.

There, in the rosy dawn light we saw women with knives dressing out whitefish for smoking. Once the fish were cleaned, the women strung them on sticks and suspended them in the cavity of an oil barrel, at the bottom of which a small pile of wood barely burned. After several hours over the smoldering wood, the silvery whitefish had become a rich golden amber.

Nuné interrupted the women at work and asked them who in town might have trout for sale. The brawny women in full-length dresses and aprons, like those in a Winslow Homer painting, stood upright still holding their knives and addressed her.

"It is late June," they said to her in Armenian, "the season for trout closes in May."

As we drove farther into the village we found the fishermen themselves. When Nuné asked them for *ishkhan* they shook their heads.

"My friends here just want to see the trout and maybe buy some for dinner," she explained to them.

"Go see Pepan," one suggested. "Pepan catches more trout than anyone. He might have some in his freezer. Go down the road a bit and turn right, or better yet, I'll take you there." The man hopped in our already crowded Niva and showed us the way.

Pepan lived in a two-story sandstone house and he came out to talk with us. He was a big man, a little less than six feet tall and broad like a bull. Despite his menacing appearance, he was a friendly man, exuding the kind of warmth derived from being powerful and secure. We handed him a bottle of cold Kilikia beer as a gesture of greeting and he embraced it in his big black arms covered with thick tawny hair. He pulled off the top of the bottle with the fingers of his plate-sized hands and the neck of it disappeared in his black beard as he chugged it down.

"Do you have any *ishkhan* we could see?" Nuné said to Pepan.

"*Ishkhan?*" Pepan said, smiling, "we have no *ishkhan* now, ha ha ha, there are not many left in the lake. It is not the season." He threw the beer bottle on the ground and it rolled into a shaded corner beside his house.

"We were told you caught *ishkhan* in your nets last night," said Nuné boldly. "My friends have come from abroad to Armenia to see the trout of Lake Sevan."

Pepan turned reluctantly toward his home and told us to follow.

He bid us sit at a small card table on a set of plastic chairs and served us each a cup of coffee with sugar. Then he pulled tray after tray of trout from a big refrigerator, his initial reluctance turning to enthusiasm about the beautiful fish. He had specimens of two of the four races of trout from the lake, *gegarkuni* and *summer bahtak,* and explained to us the differences between them.

"The *gegarkuni* has these large spots, always black, like a leopard—the *summer bahtak* has smaller and fewer spots and usually a row of red spots down the sides." Johannes and I gawked at the fish in the dim light. Before us were the drawings of trout we had studied in the old Russian paper come to life.

Pepan pulled out one specimen of *gegarkuni* that was more than two feet long. He confessed to Nuné that he had caught it in his nets early that morning. Though it had been dead for several hours, it was still a beautiful silver fish, with the big black spots like a leopard's, as Pepan had described.

Johannes and I asked if we could photograph the fish in the light, but Pepan refused. We finished our coffee and bought the best specimens to take with us, though we could not afford to pay what he was asking for the big prize fish.

"Who can afford to buy that fish for two hundred dollars?" I asked Nuné.

"Mafia," she said.

After visiting the village of Tsovagyugh our plan was to head west and south, counterclockwise, around the lake. We had marked the tributary streams we wanted to try angling in and on the way we would stop in the villages to see what the fishermen had caught in their nets.

On the lake shore near the village of Noratoos was a large group of fishmongers selling fish out of the trunks of their old sedans. Marat stopped the car and we walked along piles of smoked whitefish, dried whitefish, carp, and live crayfish. The crayfish were the same pale green color as the lake water with tinges of orange on their claws and big as a man's hand. The crayfish vendor's son heard Johannes and me speaking English and thought he'd try his out on us.

"Come to the blackboard," the boy said. "Hands behind your back. Sit down in your seat."

"This is all the English you remember from school?" Nuné asked the boy, taking his ear and twisting it. She laughed. We bought some crayfish for dinner and asked the vendor if he knew where we could find *ishkhan*.

*"Ish-khan?"* he questioned, thoroughly articulating both syllables. He waved his hand for us to follow. The boy closed the trunk of the car where the crayfish were and got into the passenger seat. His father drove away from the shore and we followed him to his house.

The man stopped the car and went inside. His wife greeted us at the door with hot coffee.

"This is pure *ishkhan,"* the man said, returning with a large object wrapped in brown paper. It was even bigger than Pepan's fish, maybe thirty inches, a kaleidoscope of blues and purples. The man stood with his hands on his waist and made an offer to Nuné in Armenian.

"He wants to sell it to you for a hundred and fifty dollars," she said.

"No, no," Johannes said, "we have enough trout to study and eat. And we have crayfish too."

As the day wore on it grew very hot and I knew the trout we had bought would spoil in the trunk of the car. The live crayfish were in a bin of water and seemed not to mind the heat.

Farther along the lake we crossed a beautiful spring-fed stream called the Tsakkar, not unlike some streams I had fished in Normandy. We turned off the road and followed it upstream to where some of its fingers percolated up from the ground. At one of the springs I saw a small bird with an amazing array of colors. It looked to be a finch of some kind, though it reminded me of a painted bunting or a warbler in a strange state of molt. Below the bird, in the water, I saw a trout holding in the current, a sight that never ceases to surprise and awaken me. I felt as if I had been in a deep slumber since the last time I had seen one.

I turned to get my fly rod, but when I returned to the pool the fish was gone, and so was the bird.

The water was warm enough that Johannes did not need to wear a wet suit to dive. He put on his mask and snorkel and slipped into the river. Both Marat and Nuné were amazed when he came up with a small fish in his net, even though it was not a trout.

"Now I see why he is so good at languages," said Nuné, "because when he dives he must ask the trout to go in his net."

We stayed the evening in a villager's home on the western shore of the lake. For dinner we boiled the crayfish we had bought and sucked their sweet meat from the vermilion shells. As I had expected, the trout had spoiled in the heat, and we fed them to a stray dog and her puppy. From our rooms we could hear the lapping of the waves, and though we were thousands of miles away, I felt as though I were on the edge of the sea.

After a breakfast of hard-boiled eggs the next morning, the four of us continued south along the coast of the lake until we reached a tributary stream called the Argichy. Where we first glimpsed it at its confluence with the lake, the Argichy looked slow and murky, but as we drove upstream on narrow secondary and tertiary roads it began to clear, like a beautiful trout stream.

The valley of the Argichy was more green and lush than other areas surrounding the lake that we had seen. Grasses and wildflowers grew along the road, and in the villages every stone home had a beautiful garden plot with potatoes, sunflowers, and hemp. Men and women were active in the fields hilling up the earth around potato plants or already harvesting the grasses from open meadows with scythes. Young girls, their hands inverted and resting on their lower backs, arched their torsos and stared at us, not seeming to be working, but keeping their watch on the solitude. The farther up the river we drove the more men we saw cutting the grasses on the open hillsides. They were sunburned and their dark naked backs were peeling.

We stopped to take a photo of the river and I watched them work. Even at a distance, I could hear the sound of their sharp scythe blades cutting the grass, like plaintive birdsongs. Occasionally, a man would stop work, pull a whetstone out from his pocket, spit on it, and make his blade keen again. The women and older men raked what grass was cut into tall stacks with wooden rakes and pitchforks.

We came to a fork in the road and stopped to ask some workers which path led to the headwaters of the stream. They were taking a break from their work, standing beside the tents where they had spent the night. One man stepped forward and offered us bread and cheese and boiled potatoes. We ate a bit and then they gave us each a shot of vodka with a tablespoon of salt in it.

"Are there trout in the river?" Nuné asked them.

"Yes," one said, "*kharmrakhait,*" which, Nuné explained, meant red-spotted; a union of *kharmish,* red, and *khait,* spots.

"That is the trout we want," Johannes said, and reminded me that in Turkey and also in nearby Azerbaijan, trout was *alabalik,* or fish with red sides.

Marat drove in the direction that the men advised for another twenty kilometers into the hills. The road became more and more treacherous and the car vibrated violently. We stopped where we could see the river off in the distance. It wound like a snake through a seemingly endless green meadow, starting somewhere in a hidden part of the mountains.

When we got down toward the stream on foot we found that the meadow was wet and soft like a cushion. The stream was forming here and all the seeps and springs we walked over were contributing to its flow. Tall grasses and wildflowers grew in the wet meadow, and though we were far above any village, men were there, harvesting and stacking it, in tall gumdrop-shaped mounds. I found it tiring to walk over the soggy ground more, I thought, than I normally should. As we approached the stream I felt a tightening in my knee. The beautiful gurgling sound of the stream through the meadow did not soothe. I was fearful something had set off the swelling again in my left knee.

I obsessively reached down to feel the knee and every time I touched it, it seemed more fluid had come into the joint. I was determined to ignore it and enjoy the beautiful place we had come into.

We came to a spot where we thought the stream looked good for fishing.

"I'll go upstream with my fly rod," I said to Johannes. "These pools look good, why don't you stay behind and dive here." He agreed to this plan and so I headed upstream, above Nuné, and Marat as well, eager to be on my own for a time.

I walked up along the banks and then through the currents of the stream itself, feeling the push of the cold water against my legs. I wanted to see if I could spook a trout and see it swim across the colored gravel of the clear pools. We had no guarantee that there were trout here at all, and I wanted to see one to build my confidence before I started fishing. I had no such luck, though, so I tied on a small caddis dry fly and began to cast it into small eddies behind boulders and beside the grassy banks.

In the shade of a large boulder where several large purple flowers were bent over and touching the water, my fly disappeared in a small swirl. I lifted up the rod and hooked the fish that had taken it. It jumped once and bent my limber rod in a pleasing arc. It sped around the pool from bank to bank, trying to find cover under the boulders until I was able to land it in the palm of my hand.

I had not noticed until then that Nuné had been following me and was watching from the tall grass on the bank. She had seen me catch and land the fish and came closer to see it.

"It is exquisite," she said. "Gosh, it is beautiful."

"Have you ever seen a live trout?" I asked.

"No."

*Kyrgyz man wearing* khalpak, *Pamir Alay, Kyrgyzstan.*

The trout was about fourteen inches, a velvety red on the sides with a marigold yellow belly and large black spots. I asked Nuné to hold the fish while I photographed it. She tied back her long black hair and held the fish for me. Then I told her she could let it go.

"Are you sure Johannes won't want to see it," she said, lowering her hands to the water. The fish flipped out of her fingers as she was thinking about it. "Oh well," she said, "there he goes."

"I'm sure Johannes will catch some of his own," I said. "I think we'll catch some more anyhow."

For the next quarter mile or so of stream, Nuné walked alongside me as I fished. I was able to experiment with different flies and enjoy the day, sit down on the bank with Nuné and describe the joy I felt while fly-fishing. She had beautiful long black hair.

"It is an art," Nuné said, watching. "It really is beautiful how the line loops around and lands on the water. And to see a trout take that fly is magical, you would never expect it." I showed her other small flies in my boxes.

"They are so delicate," she said, picking one up. "They are an expression of the predatory instinct through art. You are acting out the ritual of being a predator, but without killing the fish. That I have never heard of."

"The English perfected that," I said. "They made predation a high art."

"I guess I knew that," Nuné said. "In a fox hunt they don't even kill the fox themselves, do they? They have dogs do it."

"Sure," I said, "they just dress up in costumes and ride on beautiful horses."

"I guess a bullfight is a similar thing," she said, "but fly-fishing is more peaceful, I think."

When Nuné and I returned to Johannes and Marat we found them with a local farmer by his tent, drinking vodka. We had both been successful and shared our fishing stories. After a bit of discussion we concluded that these trout were either the progeny of a migrating population from the lake or were permanent residents.

"Maybe there are some of each," Johannes said. "Anyway, our work is done for today. We should head back to the lake so we arrive before dark."

On the treacherous ride down the road to the lake my knee continued to swell. I was seated in the back of the Niva and eventually was forced to put my leg between the two front seats because it got so bad I could not bend the joint.

"What's wrong?" Nuné asked.

"I'm not sure," I said, "but I've got this problem with my knee. I hope it goes away."

"I will pray for you," Nuné said.

In my own prayers I rarely ask for anything, but that night at a small rustic inn by the lake I did. It was a hot, uncomfortable, and still evening and I tossed and turned on the uneven mattress. I felt feverish from the anxiety that I would not get better and that I would be a burden on the trip. I lay awake and was spooked by strange sounds that seemed to echo through the town of Martuni.

In the morning the knee was worse. It throbbed like a second heart, and when I got out of bed it was excruciating to walk on. Nuné came to my room early with a cup of chamomile tea. Minutes later she returned with a bottle of fluid.

"Lay down on the bed and let me see it," she said. It appeared as though she wanted to try to treat it. I lifted up my pant leg. She could see that the swelling had increased; this I could read in the expression of concern in her face. She opened the bottle and poured what appeared to be oil on one hand and then spread it on both hands and began to massage my knee, adding more as she went. I just lay back, and after a night of no sleep I began to doze. The air in the room felt cooler and I noticed that the oil had a sweet odor.

"What is it?" I asked her in a half whisper.

"It is calendula oil." I looked up at her, where she sat on a chair by the bed, and she had no smile or expression on her face. "I think it will help you get better."

That day, Johannes went with Marat to sample a nearby stream and I rested by the lakeside and read books. Nuné stayed by my side and kept me company.

## Hamlet the Fisherman

That night a rainstorm blew over the lake and brought cooler and less hazy weather behind it. The next morning, out beyond the dark muddy surf, the water of Lake Sevan was a milky lime green. Where we stood it was sunny, but in the distance cobalt blue clouds dropped needles of rain to the water's surface.

Long after first light, but still in the morning, Marat, Nuné, Johannes, and I were on a dirt road through a semidesert land following another tributary of the lake, the Masrik River, to its source. The road was lined with Queen Anne's lace and yarrow, and looked largely devoid of human activity.

Nuné explained that the villages we passed in this valley were once occupied by Azerbaijani Muslims who fled across the nearby border with Azerbaijan. "We are still fighting them over a region east of here called Nagorno-Karabakh." Some Armenians had found opportunity in the vacant homes.

Up ahead, an old shepherd tended his flock on the barren landscape. He wore a large green poncho and rested his hands on the curved handle of his cane. Other villages deeper in the mountains had been occupied again. Women, young and old, cast suspicious glances at us from open doorways, children played in a cemetery, and the land was again cultivated, mostly with potatoes. We stopped on a bridge over the Masrik River and a man came out of his home to greet us.

"I am the mayor," he said, "welcome to my village." Nuné asked him if there were trout in the river.

"Farther up the road," he said.

"How much farther does the road go?"

"Not very far," he said. "It ends at the pass by the border."

At the end of the road, in a village nestled in a bowl-shaped valley, our car stalled and would not start. A heavy cool mist soon turned into a drizzle and then rain. An old man appeared, as old men had a way of doing, with one hand on his white beard and one behind his back.

He examined our car, peering over Marat's shoulder as he looked under the hood. The man encouraged Marat to leave the car alone and to follow him to his house.

"Come in," the man said, waving us toward his home. We probably wouldn't fish until the rain passed anyway, so we went with him.

After a short walk, we entered a small stone house. It was two stories, and we followed him up a dark steep stairwell, which I had some trouble climbing. At the top of the stairs and through a door was a large room where the old man introduced us to his wife and two daughters. He put on a gray suit jacket, sat in a chair by a table, and lit a cigarette.

My eyes migrated to a far wall of the room where a large color map of the former Soviet Union was hung on the wall. I stood up to look at it. Armenia appeared in the same shade of yellow as the Siberian tundra. I assume that the key indicated this as a dry climate with extreme temperatures. On a table below the map was a small Armenian Orthodox Bible embossed with a silver cross. There were four beds in the room, one against each wall, and a wood-burning stove in the center that radiated heat and suppressed the damp chill.

One of the old man's daughters, a middle-aged woman herself, spread out a tablecloth on the table where her father sat. We all sat around it too and were served a warm yogurt-based soup with cracked wheat and cilantro called *spas.* Then she brought a plate of *lavash* (flat bread), with sour cheese, cilantro, violet-colored mint leaves, and arugula. Through a window I could see the rain falling and puddles forming on the once-dry earth. It had grown dark outside.

The old man ate little, but lit another cigarette when a loud clap of thunder resonated in the small valley.

"Twelve of my sheep were killed last night by wolves," he told Nuné. "You can see the remains up the hill; they ripped open the bellies and ate the entrails but left the meat. I'm aiming to go up and fetch it and eat it myself." Nuné made some attempt at sympathy and let a few moments pass before she asked about trout in the Masrik River on our behalf.

"There are trout in the stream," he said, "but very few." As the rain was pouring down he told us about a giant trout his cousin had caught in Lake Sevan just weeks before. "It was the biggest one I've ever seen. He is with some seasonal fishermen, a dozen or so, on the southeast shore. They are from Tsovagyugh but they camp for the summer on the opposite side near Aregooni."

The old man encouraged us to spend the night in his home, but we declined his hospitality. The stream was blown out from the rain, it ran high and off-color, so we decided not to spend time fishing it. When we were walking out to the car I remembered that it had failed to start, but when Marat turned the key it did. Maybe it needed a rest.

Early that evening, we met with the fishermen on the eastern shore of Lake Sevan near the town of Aregooni. As the old man had told us, there were a dozen men living there, all their possessions packed in old sedans, sleeping in tents on the beach near the lapping waves. They had four boats with outboard engines and the carcasses of expired engines sat on the gravel beach.

"Hello," one said, extending his hand through a haze of cigarette smoke. He was drunk on *oghee,* a kind of grappa, and his whole body smelled of it. Nuné told them we were interested in seeing trout. A young fisherman named Hamlet invited us out in the boat to check his gill nets.

"My name is Hamlet," he said, "after Shakespeare."

He started the motor and we made our way out into the lake. The boat looked to be homemade from scrap metal and did not ride true, but meandered in the horizonless blue of the lake. When we had gone about a half mile there was no land to be seen.

The nets were hung beneath plastic bleach bottles and sunk with lead weights. Hamlet pulled the nets, which he said were set at twenty-five meters, but he caught no trout, only carp and barbel, which he did not even bother to take out of the net.

"They will be eaten by crayfish," he said. "We haven't caught a trout in two weeks. Let's go back."

The fishermen may not have had trout but they had plenty of grappa. Soon we were drunk too, sitting in their grubby midst. They all had stiff black beards, shirts caked with dried fish slime, and pants sequined with scales from carp and whitefish.

"We have to go into town, to Aregooni to find a place to spend the night," said Nuné. The stars were brilliant overhead. "No," Hamlet said, "you must stay with us." He ran over to two small tents and began to clean them out. The four fishermen who had occupied them said they preferred to sleep near the surf, next to the piles of empty bottles. Marat hesitated to stay, but the fishermen were insistent, so we slept that night on their blankets, which smelled of smoke.

The sunburned fishermen smiled in the reflection of the campfire coals as they threw fish spines and tails from their own repast into the flames. Some fell over and passed out, and only a few were up at first light to pull the nets.

My knee became so swollen that I decided I should rest it even though I had lost hope of its getting better. Nuné stayed in the car with me and talked, or I read Turgenev stories to her, while Johannes and Marat walked up streams looking for trout. I felt bad not only for myself but for Johannes. He had counted on me to be an able travel partner and now I could not accompany him or help catch fish.

"Just get better," he said when he spoke to me. I must, I thought, or it would be difficult to carry on the rest of the summer in terrain possibly more difficult than this.

Nuné rubbed calendula oil on my knee every day. She also concocted other remedies. One day she encountered some beekeepers off in the distance and walked through the tall dry grass to get a jar of comb honey. She spread the honey on *lavash* bread and had me eat it. "This will help," she said. Another day, far from the lake now, we had come to a region near Jermuk known for its therapeutic springs. Nuné bought me bottles of mineral water and had me drink them. After several days of staying off it, watching Johannes bring fish he'd caught in tributaries of the Arpa River back to the car, my knee began to improve.

By the time we had returned to Yerevan the pain and swelling were nearly gone. As a final cure, Nuné invited us to her home and cooked a piece of sturgeon meat.

"It is an ancient and strong fish from the Caspian Sea," she said. We ate the fishy-tasting fish, and I had an extra-large helping.

# Central Asia

Extending in a horizontal band between western China and Turkey, Central Asia is a largely dry and desolate region that travelers have always been eager to cross. In this sense it has never been the center of anything, except perhaps conflict. Historically, Central Asia supported a commercial network known today as the ancient Silk Road; the same roads are now used to move opium. Most were part of the Soviet Union until the early nineties, when the area was split into a half dozen or so countries with their own governments, informally called the *stans:* Turkmenistan, Tajikistan, Afghanistan, Uzbekistan, Kyrgyzstan, and Kazakhstan. When I began doing research for my trip, I learned that the *stans* were difficult to enter and politically volatile, that rebel groups ran the hill country like nineteenth-century American bandits, looking for foreigners to take hostage. These were risks a *Schwarzfischer* secretly invited.

The only person I'd met who had been to Central Asia was a fellow trout hunter from southwest Scotland named Robin Ade. He had heard about my first book on trout and contacted me while on a trip through North America. He stayed two nights at my father's house; we fished local streams, and Robin shared stories of his travels.

He had lived in Afghanistan in the early seventies, back when it was a hippie hangout, like Nepal was in the 1980s and 1990s, with much available hashish and opium. He had not been back for over twenty years and then returned on a solo expedition in 1998 to search for the easternmost native brown trout, *Salmo trutta oxianus.* This trout lived in rough territory, in streams and lakes on the border with Pakistan, some of the highest elevations for native trout in the world. It was just short of impossible for Westerners to enter Afghanistan legally at that time.

Sitting by my favorite brook trout stream in Connecticut one warm April day, Robin told me his story. "For several months in my home in the southwest highlands I grew out my beard. I bought a ticket to Islamabad, Pakistan, and from there took a jeep to Chitral. In Chitral, my beard nearly two feet long, I exchanged my Western clothes for local dress and walked to the Doruh Pass, forty-five hundred meters in elevation. From there I snuck across the border into Afghanistan as if I were a local." Robin puffed on his pipe. "It helped that I speak Farsi.

"My destination was a lake at the head of the Konkce River, a high headwater of the Amu Darya River [what the Greeks called the Oxus, which flows to the Aral Sea]. In the days following I caught many trout. The scenery was beautiful." Robin paused to puff on his pipe. "I was also traveling near a region of northern Pakistan where native people were fair skinned and blue eyed like me, so I really fit in quite well."

"Did you encounter any hostility from the people?" I asked.

"The people, no," he said. "Apart from a guy who chased me with a hatchet for trying to photograph his wife, the Afghans are warm people. But one day, an eagle swooped down and hit me from behind while I was fishing on a high ledge above the lake. I felt the wind first, and then it struck me with its talons. It nearly did do me in; I imagined it was trying to drive me off the cliff to kill me, as they do sheep and goats. Two weeks after I crossed the Doruh back into Pakistan, the Taliban fought the Pakistanis on the shore of the lake where I'd been fishing. I was very lucky with my timing. If I had stayed I would have been taken hostage or killed."

Despite the risks of being caught in the midst of a skirmish, or killed by an eagle, Robin had a positive attitude about Central Asia and its largely nomadic peoples. "No matter if they are Russian

Orthodox, Muslim, or Buddhist, the country people will treat you well. Islam especially is a religion based on hospitality. You have to tolerate inconvenience, though, and accept that travel there is unpredictable and fickle. The borders are disputed, the people have no work, and you can get unlucky."

When he left my house he wrote the following in the guest book:

"In appreciation of two days with James, not counted against our allotted span—catching magical brook trout and discussing the finer points of our esoteric researches."

Johannes and I had focused our sights on Kyrgyzstan (which borders Tajikistan on the south, Kazakhstan on the north, China on the east, and Uzbekistan on the west), a newly formed democracy risen from the rubble of the Soviet Union. From the little information we could find, Kyrgyzstan appeared to be the most stable former Soviet republic, and one of stunning alpine beauty (ninety percent of the country is mountains). It also had populations of the easternmost native brown trout, living in streams originating in the Pamir Alay Mountains.

We flew from Yerevan, over the Caspian and Aral seas, enormous bodies of water, and a large desert expanse of the 41st parallel to Almaty, Kazakhstan. Once we had landed, Johannes and I waited in Almaty all day in the terminal for Ida's arrival. We were afraid that if we left the terminal that somehow we would not be able to return. Our valid transit visas were supposed to hold us over until we got to Kyrgyzstan, but we were not allowed to stay very long in Kazakhstan.

"Ah, *mi hijo,*" Ida said when she saw me. Then she hugged me first and then Johannes. They exchanged a few words in German. I began to tell Ida about our trip to Armenia and was interrupted by Johannes.

"Okay," Johannes said, "save it for later. We have to catch the bus."

"Always business business," Ida scolded.

Silver and rounded like an old toaster, our bus to Kyrgyzstan was waiting outside the terminal in the cool night air. It took several hours to get from Almaty to the Kyrgyz border and the capital, Bishkek. It was an overnight ride on a dusty highway that smelled simultaneously of pure desert air and fuel.

We arrived in Bishkek at dawn, driving down a long avenue and past an enormous stone statue of Lenin. The city's nondescript cement structures I soon learned were a Soviet trademark, aftershocks of the Communists' stifling of creativity. They were identical to what we had seen in Yerevan, many abandoned mid-construction with pieces of rebar sticking out like fish spines. Beside the buildings, in equally drab attire, vendors sold gum and cigarettes in singles and packs and watermelon, *carpus.* Other symptoms of Soviet life were visible outside the bus window. Wooden slats had been stolen from park benches and burned during energy crises. The parks had only two kinds of trees, plane and pine, which were wild and unkempt. The faces of the people were drawn and pale like the old gray suits they wore.

The people themselves were diverse and colorful, the kind of rich diversity I had not seen since my last trip to Brooklyn. There were Slavic peoples with blond hair and blue eyes (Russians and Ukrainians), ethnic Kyrgyz with dark windburned faces, Turkic people of the western Xinjiang Province of China (called Uighur), Kazaks, Uzbeks, Dungans, and Tatars. Bishkek was a marvelous Central Asian stew and this saved it from total homogeneity.

The old silver bus dropped all passengers off at a hotel in the center of the city. From the hotel we took a taxi to the domestic airport terminal and waited for our flight to a city in the southwest of the country called Osh (Bishkek is in the north). On the border with Uzbekistan (on the ancient Silk Road), Osh is known for its political turbulence, frequent fighting, and once-great silk production.

Tired and unwashed, our luggage hanging from our hyper-extended arms, we were told that our flight to Osh, scheduled to leave at 10:30 A.M., was canceled, to be rescheduled for some undetermined time, possibly within the next twenty-four hours.

"You cannot be in a hurry if you want to travel in this place," Johannes said over beers at the airport bar. I stared out of a dirty window onto the weedy tarmac. An hour later I was restless, and Johannes and Ida were arguing, so I took a walk.

The domestic terminal was like an open warehouse with big windows on all sides. On the periphery were stone benches, where tired figures slept and slouched. Though it was excruciatingly hot, no one wore shorts or showed any skin but that on their hands and faces. That is, except for a tall blond girl wearing a red summer dress.

She could not be from here, I thought. I marveled at her perfect upright posture, her slim and astonishing figure. Then she turned and caught me staring at her. To my surprise, she picked up her bags and approached me.

"Are you on the flight to Osh?" she asked. She spoke English with a heavy Russian accent. A ray of light slanted down from the ceiling and illuminated her blond hair and her eyes, which had the amber translucency of golden raisins.

"Yes," I said, wiping sweat from my forehead, "but is there one?"

"I don't know, there's a chance they could have the flight ready as soon as sixteen hours. They are waiting for fuel."

"That's why they aren't flying, because there's no fuel?"

"Yes, it's quite common," she said, her nose twitching as she spoke. "Are you with anybody?"

I spent a second or two thinking about that and what I should say. I settled with the truth. "I am, with two friends," I said.

"I want to get back to Osh so badly. I miss my little sister."

"You live in Osh?" I said.

"Yes." Her cheeks were flush. Sweat beaded up on her temples.

"It's only six hundred kilometers distant, but the roads are so poor, and it's hot. It is hell hot. I've been in this situation before."

"You are thinking of going overland?"

"It is possible, but it takes a long time, nearly twenty hours. But on the other hand, there's no way to know if or when the plane will leave."

"Well, I hope the plane takes off by this evening," I said. "My friends and I have to be in Osh by morning to meet our driver."

"You are American, I assume," she said. "I'm Anastasiya." And she extended her hand to mine. "Where are you from exactly?"

"I'm from New York."

"Oh, New York, how wonderful. I have never been to the States, but I have traveled more than most people in Kyrgyzstan. I am returning now from Ljubljana, Slovenia. I was representing my university at an international debate, sponsored by the George Soros Foundation. "How about your friends," Anastasiya asked, "where are they from?"

"They are Austrian, a married couple. Even to me it sounds very strange, but the man is a baker by profession and his wife likes to travel. We are here to look for a rare kind of trout that lives in streams that come out of the Pamirs."

"It doesn't sound very strange to me at all," she said. "We see geologists come through and mountaineers, trout is just another reason to see Kyrgyzstan's natural beauty." Anastasiya looked at me with those transluscent golden eyes. "Aren't you concerned your friends will wonder where you are?"

"Not right now," I said. "I'm enjoying being away from them. They're not going anywhere anyway, they stay close to the beer and there is only one bar."

"Let's go sit then," Anastasiya said. I took one of her bags and followed her to a stone bench, where we sat down. I watched feverishly as she folded her long legs.

"Trout," she said, "we say *forel* in Russian."

"Does everyone speak Russian?"

"Oh no, in the countryside they speak the native Kyrgyz language. You are lucky in your country, everyone speaks English."

"Your English is very good," I said.

"Thank you. For two years actually, I had an English teacher from Missoula, Montana. He had us read a book about his home and trout fishing, you must know it. *A River Runs Through It.* I feel it is a good book because it is one I cannot forget. You close it and you're still thinking about it. At the end I was in tears."

I laughed. "I did not cry when I read it, but I did when I saw the movie."

"So you like books," she said. "Have you heard of Chingiz Aitmatov? He lives in the Chuy Valley not far from here, our only famous author. My favorite book of his would be translated *Piebald Dog Running Along the Shore.* It is about four generations of men in the same family of fishers, caught in a storm in the Bering Sea. Every day they fish within sight of the peninsula where their town is, and while they fish their dog follows them along the shore, but then they get caught in a storm that blows them beyond sight of land and they are lost, and the dog pines. There is only enough fresh drinking water for one person so the three oldest men jump into the ocean and leave the youngest with the water."

"Does the boy live?"

"I can't tell you," Anastasiya said and smiled, "you'll have to read the book."

"Here's a book for you," I said, pulling it out of my shoulder bag, "you can have it, that way I won't have to carry it anymore. It's by a German named Hermann Hesse." I handed her my copy of *Narcissus and Goldmund.*

"I will read it," she said. "Are you sure you want to part with it?"

"It's no problem," I said. "I hope you enjoy it. I'll show you where we are headed."

"Yes, do," she said.

I pulled out a map from my shoulder bag and showed Anastasiya. "We've arranged to meet a driver in Osh and drive south to this river here, the Kyzyl-Su. In this tributary, here, near this town on the Tajik border, Daraut, there should be trout."

"I'm impressed," she said, "it's not easy for foreigners to go there."

"We have the appropriate permits and everything," I said, "I just hope they work."

"Will you be staying in Osh on your return?" Anastasiya said. "I ask because I'm on vacation until August first, and if you'd like I can show you around town. There is not much to see, but I can take you to a canyon nearby where the Ak-Burra River runs through. Here," she said, pulling out some paper, "I'll give you my telephone number. You call me when you return." She wrote it down with care on a small piece of blue stationery.

"I am so looking forward to going home," she said. "I will get to see my little sister, who I always miss so much every time I go away. She's four, I'm twenty. I also have a second sister who is fourteen."

Anastasiya and I passed more time talking and then I introduced her to Johannes and Ida at the bar. She did not drink; I had a beer, and all of us (except for Johannes, who does not eat lamb) ate some *lagman,* a noodle dish with lamb and potatoes.

We decided to wait for the plane. At 2:30 A.M. the airline announced that they were combining three canceled flights that day into one jumbo jet. At this announcement Anastasiya, who had been reading, woke me (I had been sleeping on a stone bench). I found Johannes and Ida, and we began the push to get seats on the plane. Anastasiya said they often overbooked flights and you had to fight to get a seat. In the race to the plane we were up against some locals and an expedition team of Austrian mountaineers.

Outside, in the cool air of early morning, under the brilliant stars of a Central Asian night, we waited on the tarmac for the airplane door to open. Anastasiya pulled a wool sweater over her thin red summer dress. I saw several of the Austrian mountaineers looking at her and felt a tinge of jealousy. One approached her.

"Where are you from?" he asked.

"I'm from Osh," Anastasiya answered.

He seemed surprised. "And you?" he asked me.

"Osh as well," I said.

Anastasiya looked at me and laughed.

"I hope you don't mind my telling them off," I whispered to her.

## THE BASE CAMP AT LENIN MOUNTAIN

During the two-hour flight, Anastasiya slept with her head on my shoulder and I let my cheek rest on her hair. When the plane landed, our intimate moment abruptly ended. I said goodbye. Johannes, Ida, and I took a taxi to our hotel.

I welcomed having my own room, clean sheets, and a soft flat surface to lie down on. Despite the exhaustion stinging my eyelids, still I could not sleep. I was happy to listen to a rooster crowing, to think of Anastasiya, the way her hair smelled, and to feel the cold air over my face.

I went into the bathroom to pee and noticed that the toilet paper, though not suited for its intended purpose, made beautiful stationery, and looked like handmade paper. I took it from the bathroom and wrote a letter to my father. I described the rooster crowing and what I'd seen of this small city.

"I am very close to the opposite end of the earth from you, but

*Carpus, watermelon, Jalalabad, Kyrgyzstan.*

JAMES PROSEK '00

on the same parallel," I wrote. "If there is significance to this, I am too tired to elaborate on what it is."

I lay down in bed then and slept so deeply that I must have sunk below the mattress. Needless to say I felt rested by the time we had to wake to meet our driver. At the appointed time, a man named Sasha appeared outside the hotel in a green Russian jeep (called a Vilis). We met him with our bags and he packed them efficiently in the small space behind the backseat. The first stop on our way south was for provisions in the town bazaar, where Kyrgyz women sold fresh fruits and vegetables, baked goods (a flat bread called nan), and knitted items. I bought a watermelon and a bag of violet-colored plums. Johannes bought six slim bottles of vodka and a pack of cigarettes. We were leaving a fertile valley called the Ferghana for a bleak and mountainous region on the border with Tajikistan.

We drove for two hours and then Sasha stopped the jeep on a tall mountain pass from which we could view the vast desolation. Below us was a red canyon—seemingly bottomless. We were on the divide between the two major river drainages of Central Asia, traveling from the basin of the Syr Darya River into that of the Amu Darya.

"They both flow to the Aral Sea," Johannes said, eating a piece of watermelon I had cut for him. "The one difference is that the Amu Darya has native trout and the Syr Darya does not."

"Why is that?" I asked.

"I don't know," he replied.

Our first glimpse of the Kyzyl-Su River was in the village of Sary-Tasch. The river raged and roiled in a terra-cotta red color, true to its Kyrgyz name, which meant red water. At the limits of the village we were stopped by armed officials at a roadblock. Our permits and papers for travel in the Pamir Alay appeared to be in order and we were allowed to continue, now in a westerly direction, parallel to and not distant from the river. The Pamir Mountains, the highest in the world outside of the Himalayas, were visible out the left side of the jeep. Their snowcapped peaks jutted skyward, exposed, and clear of mist or cloud.

In the next village, Kashka-Su, we crossed the river on a rickety bridge. The wind was so strong it blew a spray from the turbulent water onto the jeep windows. Near the bridge, we passed a man on a horse. He wore a *khalpak,* the traditional tall white felt hat that Kyrgyz men wear.

Sasha continued to drive south, up into the hills toward the camp where we had made arrangements to stay for several days. We gained elevation, the roads became worse, and the country more desolate. We drove over clear creeks with beds of green stone that looked like jade, and by the occasional round tents (yurt) of Kyrgyz nomads. The people stared from their doorways, their goats milling outside nipping the blossoms off low-growing wildflowers. The air became colder. As we passed more yurt camps, mongrel dogs with ears clipped from fighting bit at our tires.

Finally, around a small hill, Sasha told us in his own way that we had arrived. Johannes had not expected ample accommodations but at least a shelter. What lay before us was a hollow, with a group of motley-colored tents. Behind the tents, incredibly huge, was the snow-covered peak of Lenin Mountain, 7,134 meters. We stepped from the jeep into frigid air.

A Russian man walked up to us from his red tent. He peered at us through his thick, black-rimmed glasses and zipped his puffy purple down jacket to stave off the cold.

"Hello," he said, "I am Vadim, welcome to our base camp."

Johannes, wearing a short-sleeve shirt, groped in his bags for a jacket. Vadim looked at stout and overweight Ida. She was holding her lower back from the pain she suffered on the bumpy ride. Vadim seemed confused when he saw wet suits, snorkels, and fins tied to our backpacks and probably thought to himself, what are these lunatics doing with diving gear in a dry near-Arctic climate at thirty-eight hundred meters?

"What?" Johannes said when he'd found a jacket and put it on, "there are only tents here?"

"Yes," said Vadim, matter-of-factly, "this is the base camp for expeditions to Pik-Lenina. Trips by foot to the summit begin here." Vadim looked at Ida again, his brow twisted. "You have no tent?" he said, "no sleeping bags? I think there has been a mistake. What is it you've come here for?"

"*Forel,*" Johannes said, his teeth chattering, "in the Kyzyl-Su."

"*Forel,*" said Vadim, smiling. He wore a baseball hat, which he took off then and looked at us skeptically, scratching his head through his thick black hair.

"First we must give you a tent and sleeping bag," he said. "You must realize that this is a base camp. We function on the minimal; it is not meant to be comfortable. It is thirty to forty days by foot from here to the summit of Pik-Lenina. Once they are up, they have to come back down. In 1994 we lost fifty-three people on their return, including guides, to an avalanche."

Vadim took us into the dining tent for a hot cup of coffee with cognac.

"You are here now, so we must do our best," he said, sitting down and lighting a pipe. "I apologize, but as I said, this is a base camp in rough country. First of all, to find trout you need to go down, not up."

"Yes, of course," Johannes said. He pulled out a map and, laying it on the table, showed Vadim precisely where we wanted to go.

"Where did you get such beautiful maps?" Vadim said, visibly impressed.

"In Austria," said Johannes. "We cannot find trout without these maps. Here, in this stream flowing into the village of Daraut"—he pointed—"we are looking for a population of the easternmost native brown trout. This is the stream from which the type specimen for this subspecies was named, *Salmo trutta oxianus.*"

"No problem," said Vadim, his face brightening. "You will need a truck to bring back all the trout that you will catch." He filled our empty coffee cups with ice-cold vodka. "Distilled from the tundra lichen," he said and laughed. "Climbers and fishermen are the craziest people I know."

Vadim's generosity flowed deep into the night, so despite the cold and discomfort, we slept soundly, in a state of inebriated bliss.

Through the night a mixture of snow and rain fell, but by morning the air was clear. The foothills below us were barren, smooth like velvet, and dry, with occasional abrupt rock formations thrusting from the ground. Vadim suggested that the camp cook, Natasha, the girlfriend of one of the guides on Lenin Mountain, accompany us in our trout hunt and act as a translator. She did and proved pleasant company, especially for Ida, who was feeling ill, and appeared to be in serious need of female companionship.

The river where we were headed to fish, Daraut River, flowed from a canyon in the Pamir Mountains and through the village of Daraut on the border with Tajikistan.

Sasha drove attentively on the dusty, uneven, narrow, and treacherous roads. When we had reached the crook of the valley and crossed the rickety bridge back over the Kyzyl-Su, he stopped at a small wooden shack. Fuel was sold there by the bucketful from large tanks. Sasha, with the help of two boys, poured the fuel into the twin tanks of the Vilis through a funnel. As he was doing so, a car pulled up carrying two sahibs.

They were Americans.

"What are you here for?" one asked me.

"We're looking for native trout."

"Fascinating," the other said.

"And you?"

"We are geologists."

"Don't tell me," one said, "your name is James."

"Yes," I said and looked to see if it was written on me.

"I am a fly fisherman, I have your books," the man said.

We exchanged addresses and would have talked more, but Johannes was urging me on. I got in our jeep.

"It's not every day that you meet a fan in Kyrgyzstan," I said to Johannes.

"What?" he asked.

"Nothing."

Sasha drove us up the road into the village of Daraut. We stopped by a pair of old men wearing *khalpaks,* seated in the sparse shade of a plane-tree grove.

"Are there trout in the Daraut stream?" Natasha asked them.

"Yes," one said, "but you must walk a long way." He looked at the jeep. "You can only drive so far. A rock slide has blocked the road. My son and some other men are working to clear it, but it could take weeks."

"How far should we walk?"

"About five kilometers."

We followed the stream on a dirt road and after twenty minutes came to the rock slide. Three men were working with a bulldozer to clear it.

"This is it," Natasha said. We parked beside a beautiful bank of wildflowers that grew lush and colorful by a small spring. Ida was sick and stayed behind to watch the car and our things. I sympathized with her; I felt a little dizzy myself, maybe from the altitude or the intense dry heat. She didn't like to hike very far anyway.

Johannes, Natasha, and I went by foot with food and fishing gear up the stream. The gradient was steep and the flow was fast and angry. After we had hiked a kilometer or so, we came to a mud hut. Natasha greeted the man, woman, and three children who lived there. The father offered us flat bread, honey, and butter.

We sat on some large boulders in the sun to eat. They were covered with beautiful orange lichen. A breeze moved the dry grasses.

"It looks like that stream in Portugal we fished," I said to Johannes.

"I was thinking the same thing," he said, "the Zêzere. The only difference is that you can't find a good bottle of wine around here."

I laughed a bit. "But the honey is good," I said.

I picked some yellow and purple wildflowers and pressed them into my sketchbook. With a feeling of optimism I rigged up my fly rod.

At nearly twelve thousand feet this was possibly the highest-elevation native trout habitat in the world. The ancestors of the fish in this rushing headwater stream originally swam up from the Aral Sea. They would not be easy to catch with a fly, I thought, the stream was too swift. We hiked farther to look for nice deep pools, but the stream was just a glacial-blue froth. When I started to fish I put several pieces of lead on my line to sink the fly, hoping I could get it to a fish by dropping it in an eddy behind a boulder. I fished a hundred yards of river.

I came to a small turn where the current was slower and the stream ran beneath a grassy bank. I coaxed my fly into the spot and got a strong tug. I was so excited I yanked the fish in one pull out of the water and onto the dry path behind me. I threw down my rod and leaped on the fish like a boy chasing a frog. When I caught it and beheld it, half covered in dust, I marveled at its sparse vermilion and black spots and the gentle daffodil yellow sides.

I yelled for Johannes to come.

"Good, you have caught one," he said, taking out a plastic bag to fill with water and carry the fish live back to the jeep so he could photograph it in his tank.

When we were done with the fish, we clipped its adipose fin (a small stubby fin behind the dorsal) and placed it in a vial of alcohol, preserving the tissue for DNA analysis. Then we let the small trout go.

That night after dinner we drank hot coffee with cognac. Vadim congratulated us on our catch.

"That's fine," Ida erupted. "But I am not feeling well. I want to return to Osh." Johannes thought about it, slowly drinking his coffee.

"Well," he said, "we have caught our fish, I guess it is possible to do so."

The bumpy return trip to Osh was a great strain on Ida's lower back. The air in the valley was hot and dusty, and having undergone such extremes of temperature, Ida was visibly suffering. I tried to be affectionate and sympathetic as she drank bottle after bottle of mineral water, but the person she required tenderness from was Johannes.

We had planned to explore tributaries of the Kyzyl-Su River until Saturday. It was only Wednesday and we already had returned to Osh. We were told that no flights to Bishkek were available until Monday, so unless we returned overland we had five days to kill. I knew how I wanted to spend the time.

The next morning I told the receptionist at the Osh Hotel that I wished to make a phone call. She took me into a small dark room with a telephone on a table. I wrote down the number I wished to call, and the receptionist dialed it for me and gave me the receiver.

"Anastasiya," I said when I heard a girl's voice.

"Yes?"

"It's James."

"Oh, hello, James, how are you? I can barely hear you."

"We are back early from our trip south and I thought you might like to spend some time together?"

"Sure," she said, "when?"

"Is tomorrow good?"

"Yes, tomorrow."

"We can meet at the Osh Hotel in the morning."

"Maybe at ten," she suggested. "I will bring us a lunch and we can go up in the canyon and see the falls on the river."

"Great," I said, "I'll be waiting for you in front of the hotel at ten."

* * *

At half past nine the next morning, I was waiting on the hotel steps for Anastasiya. She was dressed in shorts and a tank top, as if she were ready for a day at the beach, and she carried a small backpack. Her hair was tied back off her slender neck and the sun had worked a lovely pattern on her freckled shoulders.

"Aren't you going to bring your fishing rod?" she asked, seeing that I had nothing with me.

"Yes, if you think I should," I said.

"You can swim too."

"Are you going to swim?"

"Yes, I think I will."

"I'll bring my shorts then," I said. "I just have to go to my room to get them."

I went to my room to get my fishing rod and shorts and she followed. Anastasiya found nothing strange in coming down the dingy hall, over the old green rug to the flimsy door and into my hotel bedroom.

"Have you seen people fishing there—in the canyon?" I asked as I pulled my rod out of my bag. She stood beside me and I could almost feel her breath as she talked.

"Yes, people fish in the river," she said. "I don't see anyone there often, it's kind of my spot, or I like to think of it that way. I'm usually there alone."

I could feel my face was flush and tried to dampen the color with the back of my hand.

"Okay," I said, "I think I'm ready."

"I have a feeling you don't like my town," Anastasiya said as we were walking from the hotel to the bus terminal.

"No, I think it's beautiful," I said. In its own way it was, though it looked as though it had recently suffered a bombing. We passed a hedge of bushes with blue blossoms.

"My father planted one of these bushes at home," I said. "I don't remember what they're called."

"It's rose of Sharon," Anastasiya said, redoing her ponytail. We walked on.

"Osh is sometimes beautiful, but it's not what you would call convenient," she said. "The telephone lines are sometimes down for days."

"How was it to see your little sister the other day?" I said.

"She was kind of being a pain. She came to my bed too early my first morning home and started jumping up and down on my bed. I wanted to sleep."

The bus to the canyon was not operating, so Anastasiya and I took a taxi. Our driver was a tall Kyrgyz man who wore a starched white shirt and pleated gray pants. Anastasiya talked to him in Russian, telling him where we wished to go.

"I hope you like where we are going," she said to me, "as I told you, it is one of my favorite places."

"What is the name of the river?" I said.

"The Ak-Burra."

"What does *Ak-Burra* mean?"

"I was afraid you would ask that," Anastasiya said. In her deep Russian locution she asked the driver what it meant. "He says *Ak-Burra* is Kyrgyz for male white camel. When the male white camel is in mating season he is very furious, just like the river as it tumbles off the mountain. The falls are white as well, like the animal itself."

Anastasiya had little to no imperfections on her body. Unlike the country people, she did not have gold or copper teeth, missing digits, scars, or wandering eyes. She had not been exposed to the harsh climate, did not have to smoke cigarettes to punctuate idleness. Her toenails, which shone through her sandals, were perfect. Her legs were long and fine. Her shorts and tank top were clean and pressed.

"I have traveled abroad, more than most people in this country," she said, as she had told me before. "Most people in Kyrgyzstan who have traveled complain about the discomforts of their home when they return. A proud Kyrgyz like our driver is rare." She pointed to a sticker above his mirror, which read: Proud to be a Citizen of Kyrgyzstan. "But I think we have things here that you don't in America. Though I have not been."

It had taken about forty minutes to get to where the river rushed from the narrow canyon. We stood at the edge of a sharp decline that looked down into the Ak-Burra. Anastasiya told the driver to come back in four hours.

"It *is* an angry river," I said to Anastasiya.

The depth of the canyon was impressive, the mass of exposed rock bewildering, and still it looked as though the carved mountain was straining to hold back the power of the moving water. The only person in sight was a Kyrgyz man walking his goat.

A dry path led down to the water. Anastasiya seemed to know where she was going. When we came to the edge of the river, Anastasiya laid out a green cloth and sat on it. The river stones were small and rounded.

"We can sit here and you can fish behind that boulder where the current is broken."

"That's all right," I said.

"You've done fishing already?" she said, laughing. "Is that how you do it?"

"For now, I think," I said.

She offered me a seat on the green blanket beside her.

"This is my place," she said, "I come here after exams to calm my spirit. Do you have places like this at home?"

"Yes," I said, "I have places where I go."

"No," she said, "I meant, do you have places like this in your country, with such a river like this one?"

I did not know what to say. We did have rocky canyons and raging rivers in the States, but that's not what she wanted to hear. In my silence, she continued.

"The colors of the rock," she insisted, "the black lines coming down the red stone, and the yellows there. And look at the distant hills. I'm sure the hills in Kyrgyzstan are the largest in the world."

"We have such places but not exactly like this," I said. She seemed impatient with that answer.

"I think people would travel to see this," she said, "but people think it's dangerous to come here."

We walked up the river, barefoot in the shallows along a rock wall, until it became difficult because of the steepness of the canyon. Anastasiya was intent on getting out to a rock separated from the wall by a chute of fast water four feet wide.

"I think it's too deep," I told her, but she kept going. She found herself in the middle of the chute for a split second before turning back toward the wall again.

"A month ago you could walk out to that island," she said. "Now the water is too high."

We returned to the green blanket and Anastasiya took some food from her backpack. I had submerged a glass bottle of mineral water in the river some minutes before and now it was halfway out of the river.

"It looks like the river went down," I said.

"Yes," she said, and joked, "you know it could be the moon. What do you call that when water is moved by the moon?"

"The tide," I said. The Ak-Burra River never made it to the ocean; it flowed into the Amu Darya and then dessicated in the desert (or barely made it to the Aral Sea). It seemed sad to deprive a river of the ocean.

"The tomatoes were grown in my mother's garden," Anastasiya said, pulling them out of her backpack. "She picked them this morning. She also baked this bread. It is Russian bread. It needs a little salt."

"You have a good mother," I said.

"Nothing in this world is perfect," she said, "but my mother is perfect."

We sliced the tomatoes and ate them on the bread, and when we finished her mother's bread we tore off pieces of nan and ate it with cubes of salted pork fat.

I took a drink of mineral water and then tried fishing with a piece of pork fat for bait. I dunked the pork behind the boulder weighted with some lead. I thought about Johannes back in Osh with Ida. Were he here with us, he would have said something offensive, like, "What! You fish for Muslim fish with pork!" I would have been embarrassed and annoyed.

I lay down in the sun beside Anastasiya until I fell asleep.

She woke me, I'm not sure how long afterward, but she said it was time to hike out to meet our taxi.

I did not notice how hot it was until we left the bank of the river. The heat had a wonderful effect on the smells of the hillside flowers.

"I think you will not find smells like this anywhere else in the world," Anastasiya said, smiling. "This smell we call *poleen.*" She picked a bouquet for me as we sat on the dry ground waiting for our car and watching grasshoppers. "You can take it with you," she said, "to remember Kyrgyzstan."

After a time, we started walking down the hill toward Osh. Some way down we saw our driver washing his car by the river, where it had broken down. His hands were black from tinkering with the engine, but he had somehow managed to preserve his starched white shirt and the pleat of his pants. He told us reluctantly that we should leave him there and hitchhike back.

We walked until the sun was near the horizon. I asked Anastasiya if she'd like to join us for dinner. "We may not be back until later than that," she said. "No, I think I'm going home. If you are around tomorrow, though, we can do something else."

We walked into the starry night. Finally a car passed us and picked us up.

## The Yellow Volga, to Bishkek and Lake Issyk Kul

The next morning out of my hotel window I saw Johannes in the street talking with several drivers beside their cars, inspecting the trunk spaces and the tires. I dressed and walked downstairs.

"What's going on?" I said. I felt healthy and at home; I was thinking of Anastasiya.

"We're not waiting here any longer," Johannes said, "we have too much to see, Ida is feeling better. I think we should go overland back to Bishkek instead of waiting for our plane." He paused. "How was the time with your girlfriend?"

"It was fun," I said, though I was a little bitter.

"I think we will leave in an hour," he said.

I didn't argue. I called Anastasiya and she met us at the hotel to say good-bye. We said we would write to each other.

Johannes and I settled on a lemon-yellow Russian sedan called a Volga, driven by a small good-looking Kyrgyz man. His car looked best suited for the journey and so did he. He also gave us the best price.

About thirty kilometers outside of Osh we had our first flat tire. The driver took all our luggage out of the trunk and was searching at the bottom for his wrench and jack. The spare tire, from what we could see, was scarcely better than the old one. All of them were bald as river stones.

Once we were on the road again we stopped at the first tire-repair stand, a phenomenon of Central Asian highways. We waited in the heat, no trees in sight to find relief from the sun, while our old tire was mended and the inner tube patched.

The day was hell hot coming out of what is called the Ferghana Valley. On the steppes the wheat was weeping and even the sunflowers could not bear to face the fierce sun. Turkish music played dizzily from the car radio. We drank warm Sprite as sweat beaded up on our arms and dripped from our foreheads into our eyes. Anastasiya had warned that it was hot out on the highway between cities. I had not realized how hot.

Once the tire was fixed we traveled on, stopping to rest the car at a restaurant beyond an orchard of plums that stretched up toward the distant mountains. The restaurant was just some tables under a makeshift tent with a small brook gurgling by.

Our driver asked for some melon, which was cut into slices and set on a blue-and-white dish on a cloth of pink and white stripes. The melon had the taste of onion from the knife they cut it with, which disturbed the taste, just as the dust cloud from a passing car fogged the view of a small girl naked in the brook, splashing herself with water from a two-handled urn.

The dry heat, I thought, drained one of all superfluities.

In Jalalabad we skirted the border with Uzbekistan. Scrubby purple flowers bloomed on the roadsides. Heading north into the valley of the Naryn River as the sun set, the smooth hills were sometimes green gray and sometimes beige blue. I remembered Anastasiya telling me there were no hills as high as the hills in Kyrgyzstan. Now I believed her and no longer thought of her as a girl who had never seen the sea.

The Naryn River itself was a blue-green color. The farther up the canyon we crawled the more the river resembled a lake where it had been dammed to form the Toktogul Reservoir.

Along the edge of the reservoir the road went through a series of long tunnels with no lights in them. Many of the cars that pass through have no lights either, which makes passage dangerous. We traveled through the tunnels very slowly, having only one working headlight, and we passed other vehicles whose drivers held flashlights out their windows. Screeching trucks echoed in the tunnels like whales singing across a deep ocean. When night came it was even more dangerous outside of the tunnels as the roads were narrow and skirted the edges of cliffs.

We pulled off with our second flat tire at twilight, just beyond where a grain truck had turned over. The driver sat beside his spilled cargo smoking a cigarette. When the sun was finally down there seemed to be four times as many travelers on the bumpy road. The moon was full and lit up the water on the vast reservoir between the canyon walls. I crawled in the car when the tire was fixed and fell asleep.

Twenty-five hours after we left Osh we arrived in Bishkek (by air the distance is only about six hundred kilometers). Not once had our driver stopped to sleep and the condition of his yellow Volga sedan was not suitable for returning to Osh. He would have to either spend the money we gave him to fix the car, or sell what was left of it and take a bus home.

We checked into the Hotel Bishkek, a cement-and-steel sore on Erkindek Prospektisi. We rested in our own rooms, but just for a short while. Soon Johannes came and knocked, waking me from a deep sleep.

"We must make arrangements to travel to our next destination," he said.

This was Lake Issyk Kul in the eastern part of Kyrgyzstan near the Chinese border. The tourist agency that encouraged us to go called it "the second largest alpine lake in the world after Titicaca." *Issyk kul* in Kyrgyz means "warm lake"—because it is slightly saline and does not freeze in winter.

The same tourist agency that promised us the second largest alpine lake also found us a driver. Isa was his name, and he came by the hotel, allowing us to approve his vehicle before the trip. It was a van in good condition and we agreed to his fair price. Then he asked if he might take a friend along to keep him company on the road. Johannes hesitated in the negotiation. "Okay," he said. The van was big.

The next morning we were picked up in front of the Bishkek hotel by Isa's friend, an Uighur (from Xinjiang, China) named Marin. He spoke no English but indicated to us to come. After ten minutes we stopped by a row of apartment buildings. Isa walked toward the car, so hungover he could hardly stay on his feet.

"Wodka drink Isa," he said, sitting in the passenger seat. "Oh my head, oh mine God."

"Okay," Johannes said, shrugging his shoulders, "so what's new, our driver is drunk." I think Johannes found Isa's state was not so much disturbing as endearing. Marin drove east out of town to a straight dirt highway.

Two hours later when he had gained consciousness, Isa was friendly, sharing photos of his family. He was from Dagestan. His father named him Isa after an Italian friend, Isangeli. When he turned, you could see the profile of his large chiseled nose shaped like an eagle's beak.

Johannes instructed our pair of guides, by pointing on the map, that we wished to stay on the south shore of the lake. That shore dropped off quickly and would be good for diving. Johannes had read that there was a fishing village called Balykchy where sometimes giant trout rumored to be thirty kilos were caught. These trout were not native but had been introduced from Lake Sevan in Armenia. Since we had observed the same trout in its native habitat

*Kolya fishing a tributary of Lake Issyk Kul, Kyrgyzstan.*

we were eager to see if it retained its original markings after seventy years in a new environment.

The road to Lake Issyk Kul followed the Chu River. Many men and boys were fishing for whitefish using slender long poles baited with yellow caterpillars. As many women stood on the side of the road selling the fish dried or smoked, strung through the eyes with wire. We stopped to buy some from a lady fishmonger. She heard us speaking English.

"Do you need a translator for your trip?" she said to us.

"A translator?" asked Johannes, looking at me. "Why? We can hardly understand each other."

The sun broke through a deep layer of clouds as we caught our first glimpse of Lake Issyk Kul. Behind the far shore off in the distance were smooth ochre hills and high peaks of the Tian Shan range.

On the road to Tamga, the village where we had agreed to stay, we passed over six or seven rivers flowing north into the lake from melting snowfields in the mountains. Some were clear and others were opaque and milky, and all of them were raging.

Tamga was a village of homes once occupied seasonally by prosperous Russians. Now they were occupied year-round. Each had a garden plot with plum, peach, and cherry trees and a dog barking behind an iron gate. The area once attracted summer tourist traffic, its spas sourced by hot springs, beaches, and restaurants renowned throughout Russia. The resorts on the lake were abandoned, weeds growing through the tennis courts, rusted jungle gyms engulfed by untrimmed hedges, unkempt perennial borders revealed by an occasional bloom.

At the heart of the village, through a pair of tall metal gates, Marin drove us into an endless circuit of roads that wove around concrete buildings. In its first life, this place had been a recreation site for the Russian military, built by German POWs in World War II; in its second it had been a resort. Now it was hardly attended, a few people occupying rooms, eating prepared meals in the central dining hall, taking hikes to the lake and in the mountains.

We sat on a couch in a little room waiting to pay for our stay. A woman behind a reception desk shuffled papers for fifteen minutes, pretending to be busy, before she spoke to us. We agreed to pay six dollars a night per person for four nights. To compute this, the woman insisted on using her calculator. Then she made meticulous notations with a pencil in a book, sharpening it each time she started a new line, slipping a piece of carbon paper under every sheet so a copy would be made.

An hour passed before the transaction had been made. We put our bags in our rooms and went to the cafeteria, vast and largely vacant, to eat the dinner included in our room rate. As the five of us ate a simple borscht, a young man came over and sat with us.

"My name is Vadim," he said in good English, and spoke some cordial introduction. "What would you like to do during your stay?"

"We would like to catch trout," Johannes said.

"Ah, trout," Vadim said, testing the collar of a stiff, starched white shirt. I saw that his black pants were pleated and his black shoes shined to a mirrorlike polish. "There are many in rivers going into the lake."

"Yes?" Johannes said.

"Yes. They are beginning to come up out of the lake for spawning."

"You know a lot," said Johannes.

"I have an acquaintance who is a fisherman on the lake. It is his profession."

"Can we catch them too?" I asked.

"I think with Kolya you can."

"Then we want to meet him."

After dinner, Vadim walked us to our rooms. On the steps outside the building was another sharply dressed young man, just like

Vadim, except he wore a leather vest. Ida had recognized a clandestine tone to our transactions. "*Il mafioso,*" she whispered to me.

"Marcel," Vadim said, addressing the young man, "tell my new friends about your friend in Barskoön who is a professional fisherman."

The young man touched his head in thought, revealing that his hair was set hard like stone with grooming gel. "You can pay him to be your guide," Marcel said simply, with a high, agitated voice.

"How much?" Johannes said.

"Are you interested?" Vadim asked.

"Of course."

"Oh, I don't know," Vadim replied, turning to discuss the price with Marcel. "How about thirty dollars a day."

"Thirty?" Johannes said, putting his chin in close to his neck and looking disturbed. "How about ten."

"Twenty," Marcel said with a greasy smile.

"Fifteen."

"Eighteen."

"Okay," Johannes conceded. "We go with the fisherman."

"When?" Vadim asked.

"As soon as possible."

"Let's meet here in the morning and we can work out the details," Vadim said. "We will talk tonight with the fisherman. And one more thing: All money is to be paid to Marcel."

The opportunity to observe a man who made his living as a fisherman in Central Asia was exciting to me. It made me feel safe to sleep confidently; I felt somehow that the guide would be good.

The morning was misty through the window of my room and the lake was hidden from view. I could hear a light drizzle falling on the roof when I first awoke. As I prepared my fishing gear it began to rain harder.

"It's good weather for fishing," Johannes said when we met for breakfast, though we feared the rivers would be too high if the rain continued. Vadim met us at breakfast in perma-pressed polyester clothes and communicated the plan to us as well as to Isa and Marin. Marcel would arrive shortly with the fisherman and we would go.

I returned to my room, put on my rain jacket and rubber boots, grabbed my rod, and waited for Vadim and Marcel to arrive with the fisherman. After a few minutes, a small bruised sedan pulled up and in the backseat I saw him. The fisherman stepped out of the car and was introduced to us as Kolya.

As he shook Johannes's hand, I noticed how strong he was and how large his hands were. A thick leather jacket hung wet from his broad shoulders—had he been out fishing already?—and baggy canvas pants were tucked into his tall rubber boots. We were now quite the motley expedition team.

When we all piled into the van, I realized how much mass and speed we had accumulated, like a snowball rolling down a hill. We were now eight, bouncing westward along the lakeshore. After a short way, we turned south, up a road along a small river called the Tosorka. We had passed it the day before on our way to Tamga, and I had thought it too swift and muddy to be good for trout fishing.

When I stepped on the banks, wet from the rain and fragrant with sage and lavender, I thought, this could be a trout stream. When I stood beside the high and mad river, I was again discouraged. The fisherman assured us the fishing was good. He told Vadim that a week before he had caught a three-pound trout, and that there were bigger ones to be had.

Kolya stepped out of the car and took out his rod. It was a sturdy blue telescoping pole that extended to about nine feet. The blue of the rod was peeling and revealed an ochre-colored fiberglass. At the butt of the pole a small reel was attached, silver with a black patina. It was not a casting reel; the bait was dropped or lobbed, and Kolya wound the line by turning the spool of the reel with his finger. The line on the reel was thick, about the diameter of angel-hair pasta, and at the end was tied a hook with a chunk of lead above it and a Styrofoam float.

In his jacket pocket the fisherman carried an aluminum can with worms. He threaded a worm on the hook and dropped the rig in the eddy behind a rock and in the slower currents near the bank. He fished each possible spot carefully, holding the line in his left hand and the rod in his right. Sometimes he waited for minutes with his line in the water, circling in the eddies. He approached the better-looking holes like a cat, low to the water and slowly, not bothering to wipe the rain dripping down his face. He fished one large pool for a long time and eventually, with both hands on the pole, Kolya heaved a large silvery fish onto the bank.

It was a gorgeous trout with violet cheeks and huge black leopardlike spots, streamlined and elegant like an Atlantic salmon. Johannes and I were beside ourselves, laughing excitedly at the sight of this magnificent fish, patting the shoulder of Kolya's leather jacket. He smiled like a child. I took the fish, cleaned it off in the river, sketched it, and photographed it. It looked, as we had expected, precisely like the fish we had seen from Lake Sevan, but was a more exquisitely formed specimen.

I lay the trout on a large boulder of pink granite beside a bouquet of wildflowers that Ida had picked.

We all had a drink that night at the bar in the old military camp. Kolya drank modestly and smiled shyly, his hair, though dry, still matted down from fishing in the rain. Isa on the other hand drank furiously. He spoke loudly with his face so close to my ear that I could feel his spit on my cheek. When he wasn't speaking boisterously, he looked about the room like a nervous rabbit, his big Dagestani nose sniffing the air. I was relieved when his acute olfactory sense led him to the nearest available hooker, and he disappeared.

When Isa had gone, a calm descended on the bar and I could study and befriend our fisherman, Kolya. He had only three fingers on his left hand and with those he held his glass of vodka. More ladies came into the bar and I danced under the tall ceilings of the room.

We were so pleased with Kolya that we arranged to fish with him the next day. He arrived in the morning at the front of our residence, with several friends who wished to follow us up a mountain pass to fish for a small chublike fish called osman. Isa arrived at the van with the woman he had found the night before, but Johannes objected to her coming along. She stayed. The rain had gone and the air was dry and cool.

Isa drove away from the camp in silence, east, toward the border with China, and then south up a good gravel road along the Batshkan River. The higher we climbed, the cooler the air became, and snow still held in large masses on the hills of brittle rock. Wild poppies, orange and yellow, grew wherever there was a patch of soil. They fluttered amidst the seeps and trickles from the melting snow that cut trails and rivulets in the rock. Their petals were like delicate veined paper and they quaked in the breeze and bruised when touched.

As we neared the pass we saw higher peaks with fresh snow that looked like powdered sugar. Kolya told us through Vadim that the mountains had wild goats and that there were a few leopards. Not far from the pass, he said, he had a friend who lived in a cave for part of the year and hunted wolves and lynx with a trained eagle.

We stopped at a small bridge to fish in a milky gray pool of water. We caught many of the osman, a fish with irregular black spots, olive yellow sides, a white belly, and small whiskers, like a cross between a trout and a catfish.

It was cold enough on the pass that I needed to put on a wool hat and a sweater. Kolya's friends came over to where Johannes and I were fishing to invite us to partake of the lunch they had spread out among their catch. They had brought boiled eggs, boiled potatoes, a container with salt, and vodka. I was handed a cup of vodka and a potato. Johannes and I toasted with them. Ida missed lunch, as she was out on a walk photographing wildflowers.

After lunch, I took a walk across the treeless landscape. At every step I encountered a different type of wildflower. I picked some and pressed their delicate petals in the pages of my sketchbook. By the time I had returned to the fishing spot the Russians had gathered up their piles of osman in burlap sacks and were ready to leave.

Johannes, Ida, and I had dinner at the cafeteria in the old military compound, and afterward we sat and watched Kolya and his friends play backgammon, which they played very quickly. When the gaming was over we all went to the bar to celebrate the beautiful day.

That night there was an energy among our comrades and the drinking was aggressive. We danced as the bar filled up with people. Isa invited people to drink with us. A young woman took me by the hand and pulled me onto the dance floor. Her hair was bleached blond and she had cool dark eyes. I don't remember returning to my room, just that I ended up there, my head spinning on my pillow. I remember that I kissed her and wondered why she was not with me.

We paid Marcel, *il mafioso,* the next morning at breakfast.

"Here's my number in Bishkek if you need anything before you leave," said Vadim. "I should be back in town tomorrow, that's where I live most of the time. And let me take your address. If I find out any more about Kyrgyz trout I will write you."

We were headed away from Tamga before most of the town had awakened.

Marin was driving and Isa was passed out in the passenger seat. We stopped at a beach on the lake so we could skin dive. Isa took off his shirt and spread himself out in the sand to quiet his hangover. The water was warm enough that Johannes and I could swim and dive comfortably without wet suits.

Visibility in the lake was good, maybe ten meters, but Johannes and I saw few fish. I was happy to float on the surface staring at the bottom to quiet my own hangover, watching the geometric patterns the light made on the wavelets of sand. When we came out of the water we sat on the hot beach to dry off. I looked at Johannes beside me and saw him shaking his head.

"What?" I said, but I thought I knew.

"I think we should have given the fisherman some money."

"I was thinking the same thing, I bet he doesn't get any."

"Shit."

"Shit's right," I said, looking off across the vast lake; you could not see the other side.

"Not much we can do," he said.

Later that afternoon, on our return to Bishkek, we arrived at Issyk Ata, the last river we would fish in Kyrgyzstan before departing for Mongolia. Several boys were standing on a bridge fishing for trout. The current was very fast and they needed a lot of lead weight on their lines to keep the bait near the bottom. They had a forked stick with several dead trout strung on it.

"These are genetically pure *Salmo oxianus,*" Johannes said, lifting one of the fish, "but they are not native to this river, they were introduced by the Russians in the seventies from tributaries of the Kyzyl-Su, where we have been in the southwest."

Isa had a friend who lived in an apartment building by the river in the village of Issyk Ata and arranged lodging for us there, in a top-floor room. Another boarder, a man from the Ukraine, spoke some German.

"Where can we get a cold *kleines Bier?*" Johannes asked him.

"I'll show you," he said.

On our way to the bar across the river, Johannes asked the man if he would talk to the boys we'd seen about selling us some of their catch. The trout they had caught were bigger than the ones we had caught in the southwest, and Johannes wanted one to take back to Austria and stuff for the wall of his library. One boy sold us two, freshly caught and beautiful. He pulled them off the forked willow stick on which they had been strung and wrapped them in a pair of large ferns, carefully, like a Parisian florist might wrap a dozen tulips.

"Our work is done here," Johannes said. "I am satisfied that we have specimens, we don't need to catch them ourselves." I disagreed; I wanted to fish the river, but I didn't argue because the river was too swift for fly-fishing.

We drank several beers, cheerfully, with the Ukrainian man and then returned to the apartment for some dinner.

"We have caught and seen the trout of the Amu Darya River," Johannes said over a plate of vegetarian lagman. "It is the easternmost native brown trout." He was a little drunk and preaching. "Beyond here there are only other species of salmonid fish: taimen, char, and lenok; and grayling of course, I am very interested in the grayling."

When I painted a watercolor of the trout that night, unwrapped and lying on their bed of ferns, the act seemed to complete and satisfy an urge to collect and record. "Issyk Ata River," I wrote beside the sketch. "Headwaters of the Syr Darya River. Introduced here in the early seventies from tributaries of the Kyzyl-Su in the Amu Darya drainage."

I started a new line and wrote the month. August.

The next morning I was granted the opportunity to fish a bit myself. "We have plenty of time," I argued with Johannes, who was eager to return to Bishkek.

He accompanied me upstream anyway, up and away from the village, where I thought fewer people fished. As I had anticipated, though, the river was too swift to fish with the equipment I had.

We returned to the capital city, Bishkek, that afternoon. At dinner Johannes repeated how much he regretted not paying Kolya some money directly. Ida talked scornfully about *il mafioso.* We had become fond of the fisherman in our two days with him.

"He was a *Schwarzfischer,"* Johannes said.

We tipped our glasses and Johannes pulled a cigarette out of his shirt pocket. What we did not know was that Kolya had been hurt badly the night we left Tamga. I received a letter from Vadim when I returned home, telling me so. That he had gotten in a bad car accident the last night we drank together; they had been drunk and went off the road.

## MONGOLIA, SEARCH FOR THE HOLY GRAYLING

The 41st parallel crosses China through an arid land known as the Gobi Desert. Johannes, who had been to northwest China in search of a fish called a taimen, confirmed reports that the Chinese had eaten most of their fluvial biomass. In other words, there were few fish left. North of the Gobi Desert the land was equally arid and forbidding, but prospects for fishing seemed to improve.

Mongolia interested Johannes and me because of the diversity of its salmonid species (the scientific family that includes trout, formally, Salmonidae). There were three basic types of salmonid fishes in Mongolia: a voracious predator called a taimen, which reportedly grew to over sixty kilos; a reddish-colored and black-spotted fish with an overbite called a lenok; and several species of grayling. The grayling was a fish known in the Western world as a delicate feeder with a small mouth that rose readily to the dry fly. One of these Mongolian grayling, by contrast, was said to be a formidable predator with big, almost fanglike teeth.

The mysterious long-toothed grayling was discovered by an

English naturalist, George C. Littledale, on an excursion to Outer Mongolia in the summer of 1897. He collected a specimen of an unusual grayling-like fish purportedly from streams on the south side of the Altai Mountains, those that drain into, and dessicate in, the Gobi Desert.

Though in poor condition (Littledale had dried and salted it), the specimen he brought back to England received attention from renowned ichthyologists. It was thought by the French naturalist George Boulanger to have been the missing link between trout and grayling. He thought it peculiar enough to name it a new genus and species, *Phylogephyra altaica.* After the initial excitement waned, some disagreement arose as to the taxonomical uniqueness of the long-toothed grayling.

The Russian ichthyologist Leo Berg in his compendium of Russian fishes made the fish a synonym of the species *breverostris,* another toothy Mongolian grayling, but this without having ever seen it. Dr. Robert Behnke, who examined the specimen for his graduate thesis in the 1960s, expressed his opinion that the specimen was indeed distinct from *breverostris.* "It had larger teeth," he wrote me in a letter. Behnke granted it status as a separate species under the genus for grayling, *Thymallus altaica,* though he added it would help to have a fresh specimen to know for sure.

Dr. Behnke challenged Johannes and me to find the long-toothed grayling, which had not been seen or heard of by Western scientists for over a hundred years.

The information Johannes and I had on the long-toothed grayling, though sparse and secondhand, was all we needed in planning our trip to the earth's least-populated country (Mongolia has 2.5 million people, 2 people per square mile, and is 40 percent nomadic). I imagined the headline that would appear in the *New York Times* when we had successfully captured the fish:

*MEN END SEARCH FOR LOST GRAYLING*

Ulan Bator, Mongolia—The expedition team of Johannes Schöffmann and James Prosek made a remarkable discovery while fishing an unnamed creek on the south slope of the Altai Mountains of Outer Mongolia. At 10 A.M. on Sunday Mr. Prosek pulled a sleek silvery fish from a deep pool of the creek by the use of his fly-fishing rod. It was a grayling, a cold-water fish of the genus *Thymallus*—named so because they often smell like the herb thyme—but it was no ordinary fish.

"When we held it we saw it had long fanglike teeth on the upper maxillary," said Mr. Schöffmann, a native of Sankt Veit, Austria. Mr. Schöffmann is a passionate amateur ichthyologist who devotes his spare time to trout fishing. "After three weeks of looking we were afraid the long-toothed grayling was just a myth, but then we found this small spring-fed stream." *Continued on Page C7*

Johannes had made a contact in Mongolia through a friend in Sankt Veit. We met him at the airport in Ulan Bator. His name was Bat-Orshikh and he was president of the Union of Mongolian Journalists.

Bat-Orshikh personally escorted us from the airport into Mongolia's capital. He was an elegant man, even in the face of endless dust. He wore white gloves while driving his old BMW and spoke with his chin high even though it was the position in which one ingested the maximum amount of airborne dung. He seemed devoted to us, prepared to contribute his skills for the success of our mission.

"I am happy you have showed an interest in my country," he said while driving.

Even more than Bishkek, the city of Ulan Bator seemed imminent to return to dust. The large gray buildings were like cement-and-steel tents, erected on a windy plain, only to be pulled up and erected elsewhere. Bat-Orshikh waited while we settled in our hotel, the Zaluuchuud. The tiles in the lobby were unevenly placed, most of them loose and cracked.

We shared a meal in the hotel with our host and he ordered us each a glass of *arhi,* the Mongolian wheat vodka. We toasted and drank them down. Ida immediately requested another.

"I'll leave you here," Bat-Orshikh said, "I'm sure you will want to rest. I will come back tomorrow morning. I have made an appointment for you to meet Ajuriun Duemaa, Mongolia's only official fisheries scientist. I cannot answer your questions about fish, but Dr. Duemaa may be able to."

"What about a car?" Johannes said.

"I have arranged a van for you to take you to your first destination. It will be ready tomorrow afternoon."

Johannes, Ida, and I walked around town. We could not read any signs, not even Johannes, who could read the Cyrillic alphabet, as Mongolia had its own letters and sounds. Spoken Mongolian seemed very difficult to pronounce. If you likened French to a barking frog, Mongolian was like a burping yak. We ate dinner at a restaurant called Oscar, hung with photos of Academy Award–winning actors. It was a kind of saloon for expatriates, who were here to aid in economic development, or look for oil. The food was quite good, though almost every dish had lamb in it, which did not suit Johannes. "I don't like the look, I don't like the taste, I don't like the smell," he said.

The next morning Bat-Orshikh took us in his BMW to visit the scientist. The day was bright and clear, the shadows cast by the buildings and poplar trees stark.

Dr. Ajuriun Duemaa, whose gender I could not glean from the name, worked in one of Ulan Bator's decrepit buildings up three flights of a nearly pitch-dark stairwell. It was only when I saw Duemaa's deeply wrinkled face that I realized she was an old woman. Hunched over books and papers wearing a white lab coat on her slim frail body, she wore her long white hair in a ponytail and there was something oddly girlish in her.

We told her, through Bat-Orshikh, that we had come to

Mongolia to look for grayling on the southern slope of the Altai Mountains. Her eyes bulged in disbelief. We pulled out a map and lay it on a table. With her long pencil-thin fingers and crackling voice, she foiled our hopes of seeing the long-toothed grayling.

"I have forty-five years' experience as a fisheries scientist," she started. "There are only warm-water fishes, cyprinids, in south-slope Altai rivers," she said. "The rivers cannot support grayling, they are a cold-water fish. Where did you get your information on this fish?"

"From Dr. Behnke at Colorado State and a specimen in the British Museum. The fish is labeled as coming from the south slope of the Altai. It must exist."

"I know this Behnke, I think," Duemaa said. "I met him at a conference in Magadan [eastern Siberia]. I tell you there are no grayling in rivers of the southern Altai."

"I bet she's never left this office," Johannes whispered to me. The woman made us uncomfortable.

"Be careful, she may cast a spell on you," Ida whispered to me as we were bent over looking at the map.

"Are you sure?" Johannes asked Dr. Duemaa. He told her the story about Littledale's discovery over a hundred years before.

"I'm sure, it is not possible," she said. "Your Englishman must be mistaken."

Bat-Orshikh turned toward us and shrugged, thinking we were disappointed with the doctor's answer.

"Maybe she's right," Johannes said, showing a half smile. There was a long silence. I left the subject of the long-toothed grayling closed. I was interested in catching other fish with my fly rod, arctic grayling, lenok, and a giant landlocked troutlike fish, the taimen.

"What are the best rivers for fishing for taimen?" I asked Duemaa.

"The taimen is a red book fish. It is illegal to fish taimen," she said. "They are very scarce and almost impossible to catch." I knew this to be an untruth, because Pierre Affre had traveled to Mongolia two years before and caught over a dozen taimen on his fly rod, the biggest of which was close to two meters. Perhaps then there was a chance we would find *Thymallus altaica.*

We returned to the Zaluuchuud hotel for a late lunch with Bat-Orshikh. The fare was a small heap of rice with a dollop of ketchup, a side of sinewy mutton, and cold Mongolian beer called Chinggis Khan. The waitress sat in the kitchen listening to American pop music while we drank the Mongolian wheat vodka, *arhi,* until late afternoon when our driver, whose name was Gambatar, arrived. Gambatar sat with us in the dining room and we laid out the map on the pink tablecloths, discussing our itinerary, Bat-Orshikh acting as a translator. The next morning Johannes, Ida, and I would head into the countryside with Gambatar, southeast of Ulan Bator toward the fringe of the Gobi Desert.

We returned to Oscar restaurant for dinner and drank plenty of cold Chinggis Khan. They had a good salad of fresh lettuce and cucumber, which was very refreshing. They closed at midnight and we went to sleep in the Zaluuchuud.

At two in the morning I woke up in my room, my mouth dry. Despite the temptation, I told myself not to drink the water from the tap. Outside my room the hallway had become noisy with the voices of men and women, heavy footsteps, doors slamming, and bedsprings creeking.

Our driver, Gambatar, pulled up to the Zaluuchuud in a van the next morning. We loaded our bags in his van and started on the road to Huzirt south of Ulan Bator.

We planned to start by fishing a river called the Orhon Gol, which wound its way north through Mongolia to Lake Baikal in Russian Siberia. Orhon Gol was not a place where we might catch the long-toothed grayling (the Altai Mountains were in western

Mongolia and we were still in the east), but we had a good chance of catching arctic grayling, taimen, and lenok there.

The bumpy two-lane road to Huzirt took us through Karakorum, the town that the feared conqueror Genghis Khan had made the center of his empire in A.D. 1220. We stopped on the roadside and Gambatar prepared an early lunch of dried mutton rehydrated in a soup, Mongolia's mainstay. He had only enough firewood to bring the soup to a boil. Again, Johannes refused to eat it. Gambatar, however, gobbled up several plates.

"I don't like the smell of mutton," Johannes reminded us.

Gambatar smiled and ran his hand over his tired persimmon-shaped face and rosy windswept cheeks. The hair on the back of his head stood up where he'd slept on it. Ida had some hard sausage she'd brought from Austria and handed it to Johannes.

"You must catch us some fish to eat," Johannes said to me. "This sausage will run out and I'm not sure I will be able to catch fish by diving in these rivers."

"If you are relying on James to catch your food," Ida joked, "you may starve." Gambatar did not understand, but he continued to smile.

After several more hours of driving, Gambatar stopped the van by a massive rock outcrop in an otherwise flat and vast land. I sat in the gravel smoking a cigarette with Johannes, staring into the vastness, and then turned my face to the ground and terrorized small bugs with my spit. There was a pleasant aroma of dry grasses carried by the desert heat. I reached down to pick some powdery green leaves of a shrub that was very much like sagebrush.

At this spot by the unusual massive rocks, we saw our first camel of the trip.

"It's a true Bactrian camel," Johannes said. "You see, it has two humps."

*Bactrian camel, Gobi Desert, Mongolia.*

It stood, all feet planted on the dry earth, bracing itself in a stiff wind, its head eight feet above the ground.

Gambatar summoned us to collect ourselves with a wave of his hand. We traveled southeast on no visible road. I slept on and off, and Ida read. When we came to the river some hours later, Gambatar negotiated our stay in a *ger* camp on its banks.

We had seen plenty of *gers* in Kyrgyzstan, where they were called by the Russian name yurt, but this was the first time I had been inside one. The *ger* was a large circular tent lined with felt supported by a collapsible wooden frame and covered by canvas. It was dim inside, with four beds forming a square at the periphery, a woodstove in the center. At the opposite end from the door was a small shrine decorated with photos of the family that the *ger* belonged to.

We put our things inside. The small door, which I had to duck to get through, was ornately carved and brightly painted. On the table was a plate of dried cheese, salt tea, and fermented milk. Though the *ger* was insulated with felt and wool, you could hear the breeze blowing through the poplar trees in the grove by the river, and the gurgling currents of the Orhon Gol.

## A Perfect Day

The next morning I was awakened by Gambatar, breaking sticks and crumpling paper to feed the woodstove. It was cold enough that I could see my breath. He lit the stove and warmth from the fire quickly spread through the *ger*. The heat reddened my face, the only part of me not wrapped in thick wool blankets.

When my eyes adjusted to the light spilling in through the open door I noticed that it was snowing outside. I must be dreaming, I thought.

I put on my shoes and lumbered outside to see. A wet snow was indeed falling, accumulating lightly on the ground. It was still early and dark enough that you could see Venus near the horizon. I went back into the *ger* and unpacked my waders and fishing gear. I put on warm clothes and the waders and stuffed my fishing tackle in a small backpack. Johannes and Ida were still asleep when I left to go fishing.

Judging by the large bars of exposed gravel I walked over, the water level of the river was low. Beyond the river was a broad expanse of reddish hills. The falling snow melted on my cap and against my face. My breath curled up before me in the cold air like cigarette smoke. Not far upstream from the *ger* camp were two Mongolian men fishing a good-sized pool.

The fishermen used long rods with big spinning reels and a hook with a maggot hanging below a bobber. They cast the whole rig to the head of the pool and let it drift to the end with the current as a natural insect might. At the end of the drift they swung their rigs to the head of the pool and made the drift again.

On the bank at their feet were thirty to forty fish, mostly arctic grayling and lenok, and in their large backpacks were several dozen more. I approached the fishermen's kill like a scavenger. I had never seen a lenok and was extremely curious to have a closer look. Their sides were a mottled brick rose and they were covered with oblong black spots. Their tails were forked, their mouths more snoutish than the grayling beside them, though both were suited for bottom feeding, as their upper jaws extended over their lower. The grayling were equally beautiful, their sides covered with reddish and cerulean blue sides. Hints of buttercup yellow gleamed also, like flecks of gold in a stream bottom. The most magnificent feature of the grayling were the large sail-like dorsal fins spread like a jeweled fan. The fish flopped on the gravel until they lay dead and snow collected on their scales.

I waited there, watching, until the two fishermen reeled in their lines, filled their backpacks to overflowing, and left. Then I approached the water, rigged my fly rod, and took my time to tie on a small caddis larvae imitation. I let the pool rest a few moments and looked around in all directions. There was little to interrupt one's vision all the way to the horizon.

I cast the fly up to the head of the pool in the same manner that the fishermen had. Even though the two fishermen had taken in excess of a hundred fish from the pool, I immediately started to catch fish—first a grayling and then a lenok—and was elated. My enthusiasm warmed me as I began to see the redness of the sun envelop the clouds and the reddish hills.

Back at camp I entered the darkness of the *ger* again. I warmed my hands on the dry heat from the stove. Pungent wood smoke and that from burning dung tickled my nostrils. Johannes looked up from sleep, his eyes glinting in the light that spilled in from where the stovepipe exited. I held two fish close enough to his face that he might smell them.

"Oh, you have caught a lenok," he called, jumping out of the warm blankets and groping on the floor for his glasses. "Is it live? I must photograph it in the tank!"

"It was snowing this morning," I said, "where were you?"

"It must be cold out, then." Johannes found his camera and was about to walk outside when he realized he'd forgotten to put his pants on. We walked outside to observe the one grayling and one lenok I had kept. He photographed them as they lay dead in the low grass outside the *ger*.

"We need live ones," Johannes said. "Can you catch more?"

"I'm pretty sure I can," I said. "The river is full of them."

A young woman who tended the *ger* camp prepared our breakfast, hunks of bread and dry hard cheese and a bowl of mutton and potato soup with noodles. The woman also cleaned and cooked the fish I had kept, which made Johannes happy. The grayling was sweet and fine textured, more savory than the lenok, whose flesh was somewhat grainy.

The snow had long stopped and now the sky opened up and a hot sun beamed. The air became warm and I took off most of the layers of clothes I had worn that morning. The sky was alive with cottony clouds dancing at the fringe of the racing front. All signs forecasted a perfect day.

After breakfast, I took Johannes upstream, past the pool where I had caught the fish that morning. Ida came with us part of the way and then returned to camp because her feet were hurting.

The farther we walked the greener and more lush the grasses became. Here and there was a stand of blooming irises, the grass around them clipped neatly to their bases by brown sheep grazing in the meadows.

A mile upstream of camp I began casting. It was not long before I hooked and landed a lenok, which Johannes put in a plastic bag of water so he could bring it to camp and photograph it alive in his tank.

"Hey, look over there," Johannes said, pointing to the opposite bank of the river. A single boy stood watching us, and after a few moments he waded through a shallow riffle of the cold water to our side.

The boy followed us as we continued past a small pen of piebald goats. He was small and had trouble walking because his big black cowboy boots were twice his proper size. His hands were hidden in the sleeves of an oversized man's jacket, which swung like an elephant's trunk on either side of his hips. A red rag held this robe together around his waist and his short-cropped hair was mostly concealed by a fur cap, partially shadowing his parched and windswept cheeks.

The three of us came to a long pool, the kind an Atlantic salmon fisherman would recognize as perfect, a steady even riffle tailing out to a deep-throated hole. There were so many fish I began catching one on every cast. Johannes stood beside me for some time looking impatient.

"Okay," he said, "you have caught enough? I have one of each for photographing and an extra for dinner. Shouldn't we go back to the camp and check on Ida?"

"You can," I retorted. "I'm going to stay awhile. We're always running around sampling all these rivers at your frantic pace. I've got this one all to myself now, there's plenty of fish, and I'm not leaving." I realized I was just short of stomping my foot like a reluctant child. Both Johannes and the boy were staring at me. "It's a perfect day," I said calmly.

"Okay then," Johannes said. "I'll go back."

I and the strangely dressed boy continued across the barren land, stopping as I fished pools that looked promising. The boy picked up a long slender stem from a scrublike bush to mimick my casting, then he chased some grazing horses with it. Several times when I hooked a fish I handed the rod to the boy and he reeled it in. Occasionally he would rest nearby like a fragile prince, lying flat on the ground beneath the incessant wind.

After a while I said to the boy, "That looks like the end of it." The river had become wider and flatter and there were no longer any deep pools holding fish.

He began to sing.

I sang also and the boy seemed amused. Then I whistled and he wrinkled his brow to show that he was annoyed. He indicated he wished to challenge me to a race, so we began running. As we ran, we came closer and closer to the spot where we had met. I stopped running when it looked like he might trip from his oversize boots.

"I'm out of breath," I said to the boy, panting. He began to walk away from me.

When we were some distance apart and he was just an object

kicking dust, I looked back and saw him waving. I had no idea where he was going.

Back in the *ger,* Ida was reading a book, a fire was burning in the woodstove. She didn't say much to me, nor did Johannes.

At dinner we spoke a lot, but mostly about how good the fish tasted. We ate plates of fried grayling; the flesh was golden orange, sweet and delicious. For dessert we were given some yogurt with wild berries.

The next day Gambatar drove us over the open and roadless land. He proved a helpful and amiable travel companion. My only complaint was that he played the same tape of Boney M songs over and over on the radio. Shortly, I knew the lyrics to all of them: "Ra Ra Rasputin, lover of the Russian queen. Ra Ra Rasputin, Russia's greatest love machine."

At another section of the Orhon Gol, we came upon some men fishing with hand lines baited with whole dead lemmings.

"I'd like to see the fish that eats that," Johannes said.

"Maybe we should dive in the pool and see what's down there. It should be taimen."

"I think we'll wait until they leave," Johannes said, but the fishermen stayed longer than we did.

Two days later we fished some rivers flowing to the Gobi Desert that could have been inhabited by Littledale's elusive long-toothed grayling. The most promising of these watersheds was Ongijn Gol, a cold and clear stream, fertile with aquatic insects and a suitable habitat for grayling. But our efforts produced only a few diminutive loach, which Johannes caught in a hand net.

Ida seemed more weary than usual of our fish concerns and incessant obsession with the long-toothed grayling. Her stout and corpulent body was taxed riding over very rough terrain. She held her lower back with both hands, her face sweating and her mascara running. I noticed after a while that tears were streaming down her face and she leaned over to me, as we were sitting beside each other in the backseat, and spoke in my ear.

"I am deeply sad," she said to me. "Johannes is not an easy man to be with."

I put my hand on Ida's shoulder. There was enough noise from the engine that Johannes could not hear us.

"What can I do, Ida?" I said. "I wish I could do something. He is a crazy man, but what you hate about him makes him interesting to me."

"There is no question he is interesting," Ida said defensively. "Johannes is very smart, you know. He is incredible with languages, he's just not an easy man."

Johannes, Gambatar, and I spent the early evening in front of a hotel in the village of Arvajheer, smoking cigarettes and drinking beer, while Ida rested her back in the room we'd checked into. My radar detected the voice of an American.

He wore a turtleneck that read Denver Broncos on the collar. Johannes and I heard every word of what he was saying to the small Mongolian man with whom he spoke. After a time, he spotted us as Westerners and bought us a drink. He was from the International Republican Institute, visiting local officials and consulting them on how to run a proper democracy. It seemed the locals of Arvajheer liked him. They were throwing him a party.

"What's the primary occupation of Mongolians?" I asked him over our drinks.

"Herders, they're nomads," he said. "There's no economy."

"That seems all right," I said.

"You can't have a democracy without capitalism," he said.

"What about tourism?" I asked.

"There's stuff to see but no infrastructure. The roads suck, I'm sure you've seen—you can't get anywhere if you don't have a chopper. Have you taken any of the domestic flights?"

"We are tomorrow," Johannes said. "We're flying to the western Altai, to the town of Hovd."

"God bless you," he said, laughing. "Keep in mind when you're boarding that they dropped seven out of eleven planes in the past three weeks, four emergency landings and three crash landings. The parts are made in New Jersey, assembled in China, and Miat Airlines flies them, that's a bad recipe. God bless you," he said again. "It's okay, they're being extra careful this week, there's a lot of foreign dignitaries in town."

Ida never showed up for dinner—she must have fallen asleep in the hotel room—so Johannes and I and Gambatar forgot to eat as well. Gambatar went to sleep in the van and Johannes and I continued to drink until two in the morning. Then we returned to our room with its two beds. I slept on the floor between them.

I was about to fall asleep when I heard Ida, first whimpering, then sobbing. Johannes did not cross the room to console her; he pretended he was asleep. I decided to leave the room.

I sat in the hallway listening. Ida began to scream at Johannes in German. He did not reply. About ten minutes later she came out of the room and found me sitting on the floor in the dark hall. She had brought her cigarettes and wanted me to share the time with her.

*"Ven conmigo,"* she said, still crying.

I gave her my hand and led her down the marble stairs, chipped and decrepit, until we were outside.

"Johannes has a girlfriend," she said. "He doesn't like me because I'm fat. He likes thin women."

"Oh, Ida," I said, "you're beautiful. What do you need Johannes for? You have your children."

"We have been married twenty-two years but Hannes doesn't care about anything except trout. He gets worse every year. It's a craziness."

Ida wet my shoulder with her tears. "He works when I sleep. And the rest of the time he spends in the bar."

Somewhere in the night dogs were barking; it seemed there were hundreds. The sky was clear but I could see lightning beyond the far hills.

Ida was still sobbing when Johannes came down from the room and sat with us there, on the steps outside the cement building. He lit a cigarette but did not speak. He was on the other side of Ida from me.

"Did you see it?" he said after a time.

"What?" I said.

"The shooting star."

"No," I said.

Ida was silent.

## The Altai

The next afternoon, back in Ulan Bator, we boarded a domestic flight to Hovd, a village at the foot of the Altai Mountains. As with other planes we had taken within poor countries, there was no system to seating and the flight was severely overbooked. An old woman half my size pushed me at the midriff to get by, others just sat in the aisle. Unless you had money to rent a helicopter, though, there was no other way to get across the country in a reasonable amount of time.

Bat-Orshikh, our journalist liaison, arranged for a woman to meet us at the airport in Hovd. That woman, named Singsee, and a man named Bolt, helped us find our bags when we arrived, and then escorted us to a hotel in town.

Singsee took our passports, which was disconcerting, but she said she must as they were to be held by the local police as long as we were in the Altai region. Later we found ourselves deep in negotiations with the hotel manager for the price of a room.

"Thirty dollars for a room with two cots and no running water is way too high," Johannes said. It was not that we couldn't afford it, it was simply that they were charging too much for what we were getting.

"There's nothing I hate more than the feeling of being cheated," Ida said. Johannes negotiated them down to fifteen.

Our destination was a river that connected two huge lakes called Khar Us Nuur, the Black Water Lakes. Between Hovd and the lakes was open land scarred only by a few indentations from jeep tracks.

Our driver was drunk when he picked us up at the hotel.

"I can smell alcohol on his breath," Johannes said to me as we were about to depart.

Singsee stood beside us. "And now," she said, "it's *showtime,*" which was her way of telling Westerners that we had to pay for the jeep in advance.

"These are awful people," I said to Johannes.

There was no danger in our driver being drunk, though, and no danger of going off the roads, because there were none. What bothered us more was the condition of the car.

"The tires are as bald as Johannes's head," Ida observed. When Bolt, the man who'd met us at the airport, climbed in the jeep with us, we had no energy left to resist.

A boy in a cowboy hat standing outside the hotel noted our disappointment.

"I could take you in the mountains to look for snow leopard," he said. "I have a good vehicle."

"We're not interested in snow leopards," Johannes said.

The boy reached down to pick a piece of dry grass and put it in his mouth.

Twenty kilometers outside of Hovd we got our first flat tire. The driver put on the spare. Thirty more kilometers over dry earth we had our second flat. Both blown tires had gashes in them three inches long.

I observed then the true meaning of resourcefulness and ingenuity. The driver cut a piece of rubber from one of the tires with his pocketknife and put the patch on the inside of the other tire to cover the hole. Then he took a new inner tube, put it inside the tire, and filled it with air, and the air pressure held the patch in place. I cringed when I saw that the new inner tube was also patched, but his quick fix worked.

"The ultimate problem," Johannes said, "is that we don't have enough time and they have too much."

Along the way to the Black Water Lakes the driver and Bolt stopped at several *gers* in the desert to visit friends. We knew for certain now that we were on *their* time. From Hovd we went to Buyant, Myangol, Dörgön, and eventually we made it to the Chonokharaykh River between the lakes.

Beside the river was a group of *gers* where we were offered warm milk and hard dry cheese by a man who had unusually bright white teeth (for a Mongolian). Good-teeth, as we came to call him, understood when we told him that we wanted to fish for the native grayling, *Thymallus breverostris.* When we had done eating, Good-teeth joined us in the jeep.

We drove down to the bank of the wide river toward a big cliff, half brick red and half black clay. On the opposite bank camels were grazing and downstream you could see where the river spilled into one of the Black Water Lakes.

When we parked, Good-teeth got out and pointed to where he

thought I should cast. I rigged up my fly rod and he looked at it, shaking his head. He made with his hands that there were big fish and that my rod was too small.

The river was dark and deeply stained like black tea. I put on a white streamer fly that would be visible in the water and on the third cast I pulled out a bright silver fish.

I managed to get it up on the bank on a bed of low stubby grass. Johannes leaped on it, bracing it between his knees.

"Oh my gosh," he said, "it's *Thymallus breverostris!* I am almost certain there has never been a color photograph taken of this fish."

It was not typical in our minds of what a grayling should look like. The European and American grayling had a large colorful dorsal fin, a small mouth, and brick-and-yellow sides. This fish had a large mouth like a trout and a set of formidable teeth and larger silvery scales. It was easy to see how a biologist like Boulanger, studying Littledale's specimen, would call it the missing link between trout and grayling.

The fish matched descriptions of Littledale's fish but we were not on the southern slope of the Altai on a stream flowing south to the Gobi Desert. We were fishing in a north-flowing Siberian watershed.

"Maybe Littledale had discovered the fish in a river draining north and had been disoriented by the snakey meandering of the rivers," I said. "It's easy to be disoriented out here. I think that *altaicus* was just a local variety of this fish."

"Yes," Johannes said, "perhaps Dr. Duemaa was right and there are no grayling on south-slope rivers. In order to find out for sure, though, we need an entire summer, maybe two, and a helicopter. So then, we have an excuse to return to Mongolia."

When it was dark we returned to the village.

Bolt walked into the *ger* first and lit a candle. Good-teeth lit a small gas stove and put a woklike pot on it filled with four ladlefuls of water from a jug. He handed a lump of dried mutton to our driver, who cut it up in pieces with his pocketknife. The driver added shoelace-sized pieces of dried mutton to the pot of water.

Meanwhile, Good-teeth got a second pot of water boiling and put a lump of something that looked like an owl turd in it.

"Mongolian coffee," he said and smiled, showing his white teeth.

He served the hot drink with a dollop of butter in it. It tasted like Louisiana chicory coffee. A woman entered the small space. She lit a second candle, illuminating a young and beautiful face. Kneeling on the earthen floor, she slowly added water to a bowl of flour, kneading it into a dough, until it was the right consistency to roll out and cut into strips. Good-teeth pointed to the girl and indicated that these noodles were called *sah*. They were added to the pot with the mutton and Good-teeth stirred them.

Smoking and talking, we waited for the soup, and when it was ready we slurped it out of small bowls. Loud slurping was proper *ger* etiquette, as was accepting all offerings with your right hand; the left was sinister. Glasses of wheat vodka were passed around, but I made only the first round before I fell asleep.

Having successfully caught our grayling, we returned to Hovd the next morning. On the way we had our fourth and last flat tire. We at last were stuck on the wide and desolate windswept earth. Bolt and our driver puffed continuously on packaged cigarettes until they ran out, at which time they started rolling loose tobacco in pieces of newspaper. How long were they prepared to sit there? Several hours passed, then a motorcycle rode by and stopped to pick up our driver in his sidecar. Several hours after that we spotted our driver on the horizon, rolling a tire in front of him.

We arrived in Hovd late that evening, and the following day we flew back to Ulan Bator. We had enough time left in Mongolia to spend a day in town and go on a short trip east of the city to fish the headwaters of the Amur River (one of the biggest rivers in Siberia).

Compared to Hovd, Ulan Bator seemed the pinnacle of the civilized world. Back in the comfort of the Zaluuchuud hotel we had clean linens, firm beds, and running water. We contacted our journalist friend Bat-Orshikh, and he arranged for the return of our faithful driver, Gambatar, to take us east.

Bat-Orshikh recommended a trip to the Natural History Museum of Mongolia.

"You may see some interesting fishes there."

After seeing it, I thought, there is anthropological value in touring a country's Natural History Museum. Paris's was just strange, the statue of a giant orangutan strangling a young man in the lobby, the obsessive cataloging of rooms of bones, the skeleton of a human cyclops and other birth deformities in glass cases. New York's was not so edgy but de-gritted for the American imagination—magnificent dioramas with elegantly rendered backdrop paintings, clean and well lighted. Ulan Bator's was a natural historian's basement. At the door we were given a flashlight and apologies that the electricity was not working. Some of the rooms had no light at all and the eyes of poorly taxidermied snow cats, the stitches of the craftsman's art showing in places, gleamed yellow in our handheld lights. The experience was altogether odd and fantastic; I felt as if I had to protect myself or light a fire in a corner to keep the beasts from the mouth of the cave. There were fishes, yes, but they looked more like giant raisins with teeth.

The next day we headed east with Gambatar in a cold rain to a river called Kherlen Gol. The river ran through a thick evergreen forest with lush undergrowth, a landscape unlike any we'd seen in Mongolia. We settled in a *ger* camp spread over a large meadow.

During the day Johannes and I dove and fished in the cold river. The water was clear and there were many lenok for me to catch with my fly rod. Gambatar joined us by the river and cast my spinning rod. It was still summer, late August, but I could feel the coolness of autumn, reminding me of home.

The conversation at dinner after some vodka was clever and playful, a conglomeration of words that Johannes, Ida, and I had picked up on our travels together. Ida and I were drunk and she held me closely to her side. She kissed me over and over on the cheeks and called me her son. I had to come visit her in Sankt Veit, she demanded. "Don't leave me with Hannes," she said in front of him. "You understand, you understand." Gambatar did not understand. He lit a cigarette and offered one to each of us, smiling.

Well liquored, the four of us went to sleep in the *ger* under thick wool blankets with the woodstove piping hot. Hours later, in the cold morning, Gambatar woke me when he lit the fire again, which had gone out. Then he went back to bed.

While they were still asleep I snuck out of the *ger* to the river with my spinning rod and on the second cast I hooked a huge fish. Upstream it went, down it went. It jumped, leaving the water completely, and touched its head to its tail. It embodied François's sculpture *le grand bécard vainqueur,* the one that should be in America waiting for me. The one I will watch this winter, I thought, as snow falls outside my window. Well, here it is, a line and a hook and you try to get the fish to take the hook. And there he is finally at the end of my line, a monster running up and down the river, myself and it; it weighing probably half my weight and what shall I do, what shall I do?

Ida smiled deeply at me, her eyes slimming to nearly shut. Any amount of effort it took to speak to her or Johannes was smoothed by the beer we drank in the airport bar as they waited with me for my flight to depart.

I sat beside Ida and she scratched my back, massaged my shoulders, which eased a nervous excitement I felt at the prospect of being alone.

"Don't leave me with Hannes," Ida whispered to me in the broken form of Spanish we had come to speak, and then laughed.

"But you will be back in Sankt Veit soon," Johannes said. "There is still Algeria to search for trout, and I have new information on a stream in Sicily." A minute or two passed. One or two cigarettes were smoked and Johannes spoke again. "So, *amigo,* I think it's time you catch your plane." We stood up. I hugged Ida and shook Johannes's hand.

*Mongolian boy holding arctic grayling, Orhon Gol, Mongolia.*

# Part IV

# The Transit Lounge in Seoul, Korea

They say that the time spent fishing is not deducted from a person's days on earth. When you consider that stress is known to exacerbate every illness from the flu to schizophrenia, this does not seem so far off; that the gentle art of fishing is not only a pastime, but a tonic. In the seventeenth century, when the average life expectancy of a man was forty, the father of modern angling and author of the *Compleat Angler,* Izaak Walton, lived to be ninety years old.

I flew from Ulan Bator to Seoul, Korea, and was now alone, in transit, examining my passion for fishing within the context of my life.

I walked slowly by the duty-free shops in the Seoul airport, their neatly stacked wares, the flight attendants in pristine uniforms. I sat in the transit lounge across from a beautiful young woman who was reading a Russian crime novel.

She looked up and saw me staring at her.

"Hello, I am Katya," she said, "in English my name is Kate."

"I am James," I said.

"You want a smoke? I was going to the smoking lounge."

"Sure," I said.

"And where were you?" she said as we walked. With her heels she was nearly my height.

"I was in Mongolia with two friends."

"Ooh, Mongolia," she said, half mockingly, "you are a wild man! And what were you doing in Mongolia?"

"My friends and I were fishing for a rare kind of fish."

"Did you find it?"

"No."

"But that's what keeps you going, isn't it?" She laughed. "You seem a rare fish yourself. I have been fishing just yesterday near my home in Vladivostok."

"For what?" I said, sitting next to her in the lounge full of smoke.

"Oh, the fish with two eyes on the same side of the head."

"A flounder," I said. Katya took out a cigarette with her slender fingers.

"That's right," she said and lit her cigarette. "How old do you think I am?"

"Twenty-two," I said.

"No, I am eighteen."

"Where are you headed?" I asked.

"Australia," Katya said, putting out her cigarette. "Would you like to eat something with me?"

"I would," I said, "but I have no cash."

"Okay, then I will buy the American his lunch."

We walked side by side to the transit lounge cafeteria.

"Do you have a girlfriend?" Katya asked.

"No."

"Why not—you are a priest maybe?"

"I'm sure I am not a priest," I said.

"You have a girlfriend, then?"

"Sure."

While we were waiting in line for food at the cafeteria, a Chinese man asked Katya and me if we would like to join him at his table. We introduced ourselves when we sat down.

"You are not married, then?" the Chinese man asked. "Oh, but you would make a nice couple."

The Chinese man was impressed that I ate my noodles with chopsticks and tried to teach Katya how to use them.

"Do you know who I am?" he said after a bit. "My maternal grandfather was the first prime minister of China."

His name was Mr. Chu and he lived in Allentown, Pennsylvania. Chu liked to talk and his words intrigued me, but Katya was bored with him.

"These transit areas are timeless, aren't they?" Chu philosophized. I was thinking about how we were all going different ways.

# Kamchatka, the Russian Far East

I was flying from Seoul to Petropavlovsk, the largest city on the Kamchatka Peninsula in the Russian Far East. There I would attend a meeting of international fish biologists, the Arctic Char Fanatics, an organization founded by a Swedish scientist, Johan Hammar, for the purpose of studying arctic char. The forty or so members of the Fanatics (there were at least two from various northern countries) met once every two years in a different near-arctic location to talk about char, deliver papers, fish, and consume large amounts of alcohol. I was invited along by one of the members, a biologist from the state of Maine, my friend Fred Kircheis.

When I landed in the small airport on the southern tip of the Kamchatka Peninsula, I saw Fred, for the first time in two years. He and several other members of the society had just arrived from Alaska.

"The last time James and I saw each other," Fred said to another colleague, "was on a short expedition in search of blueback trout in Rainbow Lake, Maine. We spent a week in a small cabin taking samples with gill nets." I was delighted to hear his Maine accent again.

Of the thirty-two scientists on the trip representing eleven countries—Sweden, Finland, Norway, Scotland, America, Canada, Japan, France, Switzerland, Austria, and Germany—I was the only one whose bags did not arrive. They had never, in fact, left the plane, which was on its way to Anchorage and would not return until the next day.

Our Russian hosts, a handful of ichthyologists from Magadan, Vladivostok, and Moscow (which included two members of the Duma, or Russian parliament), greeted us at the airport and assured me that someone would stay behind with me until my bags arrived. The rest of the party would move north up the peninsula to the mouth of the Kamchatka River and occupy a rustic research facility on a tributary called the Raduga.

Being left behind I began to see as a blessing. I was alone with a pair of Russians, a tall young man named Igor and a young woman meant to be a translator named Sasha. As we waited in the airport for papers to be processed, Igor offered me vodka and we enjoyed some laughs together. He and Sasha gave me a tour of the port city, which differed from other Soviet cities I'd seen in that it was on the sea. Because of the influence of salt water it also had another color in its palette of earth tones, the brilliant orange of rust.

I was put up in a small boardinghouse with a view of the harbor. The ships moored there were massive, like prehistoric whales. The sea was cold but the weed and rust smells were somehow inviting.

Early the next morning Igor and Sasha picked me up, but a third Russian was driving, a girl named Olga. My bags had arrived and Igor had already fetched them. We were heading north to join our party.

The Fanatics were at a boardinghouse two hours north of Petropavlovsk. I was sitting in the backseat beside Sasha and briefly I caught a glimpse of Olga's eyes in her rearview mirror. They were strikingly green and her reddish hair was as wild as if an electric current ran through her. We rendezvoused with the Fanatics for a late breakfast and afterward were on the road, in a caravan of four vehicles heading north on the dusty road. In the rearrangement of people and vehicles, I had been separated from Olga.

The first part of the two-day journey to the research station was across a flat green land. In the distance was the tall peak of a massive volcano, perfectly formed like an isosceles triangle.

"That is Mount Cruchevskaya," Igor said, pointing. "It is thirty-nine hundred meters high."

The van was cramped and hot but I didn't mind because I was

forced to sit close to the Russians and I could study them. I was with Sasha and Igor, and another young woman, Oxana, and our leader, Dr. Glubakovsky, who looked a bit like Gorbachev. Each had some mystical oddness to his or her appearance. Oxana was like a faerie with long delicate ears. Sasha had one brown eye and one blue and the blue never looked precisely at you. Other Russians in our party looked like wolves or foxes.

Our caravan drove on through the night, stopping at a guest house in the village of Milkowa for dinner. Vodka flowed like small clear brooks, babbling from tall bottles into teacups. The scientists—thirty-nine men, and one woman from Wisconsin—conversed loudly, and when the magic took effect, they danced on the unfinished wood floors.

The next morning, awake in a location where few could remember arriving, we drank shots of vodka with breakfast and bathed in hot springs down a dirt road from the guest house. On the way back from the springs wrapped in towels, we spotted a bear in a cage.

"Typical of the Russians to put a bear in a cage," said Fred DeCicco, an ichthyologist from Alaska.

"Ten years ago an American walking in Kamchatka would have been shot, or caged like this," said Markuu, a Finnish scientist with a white beard.

Half drunk before noon, my Russian companions and I joked in the van. The driver was moving fast over dry sand and the roadside trees were white with dust. As the day warmed I took off my jacket and wrapped it around my head. The Russians laughed at this and called me Arafat.

The driver stopped on the road by a patch of wild blueberries. He offered me a Bulgarian cigarette. The blueberries were flavorful and their leaves were mottled green and bright red. Autumn was taking hold. The rest of the caravan stopped behind our van and the scientists fanned across the berry patch, identifying everything they saw. Markuu pissed on a plant and said, "Hey, what's this?" Fred inched up the hillsides, looking. There was a tall red mushroom that none of them had seen before, a purple beetle, a yellow feather.

Beyond the blueberry patch the caravan stopped on the bank of the wide Kamchatka River to await a ferry. The scientists searched along the river's high-water mark, their heads bent, lifting dead sticklebacks from the dried mud where the water had receded and left them stranded. Amidst the sticklebacks was a dizzying pattern of overlapping bird prints, accentuated by the sun-cracked mud. They raped the soil with their eyes wishing to identify it all. A rusted sawblade, what species? *Sawbladus crosscutticus rusticus.*

Humans and vehicles had crossed the cold silty river, and then down the road, the caravan stopped at an open market. We tasted some smoked sockeye salmon and drank a liter of vodka to the health of the vendor's daughter.

"Fuck that the economy is dead," Igor said, toasting, "fuck that the ruble is worthless."

We drove another forty miles over sand and through mud to the Kamchatka River again, which had snaked around to meet us a second time. Three iron boats shifted between off-kilter pilings and one rusted man stood in each waiting to take us across. This time the vehicles stayed. We would travel thirty kilometers by boat down the Kamchatka River, and then up the Raduga River to the research station.

Dr. Glubakovsky, our host and Duma member (deputy chairman of the Committee of Education and Science), stood on the bank of the wide river. "Before this day," he said, "only two Westerners have ever been to this remote station, the president of Finland and the head of the East German KGB. I helped build this station myself as a young biologist in the 1970s. I hope you feel welcome there."

We pushed off the muddy bank and our boat headed across the vast river. It was nearing twilight and you could see the opposite bank, but then a fog set in and our engine caught on fire. The Russian operating our boat filled a bucket with water and threw it on the fire. It took a second and third bucket to put it out, then he slowly approached it as if it were a meteor that had struck the earth.

The passengers, Norwegians and Scots, a Finn, and one Japanese man from Grenoble named Yoichi Machino, looked beyond the gunwale and saw nothing, except a deep blue sky that was growing black. We were drifting down the river toward its mouth and the Bering Sea.

Meanwhile the fog grew thicker. Markuu, the Finn, grabbed a bottle of vodka. We passed it around and he started to sing. Soon we were all singing, easy sing-alongs, *What shall we do with the drunken sailor,* melancholy songs that sounded sublime in our desperate state in the dark and fog.

Before long we saw a floodlight emerging from grayness, and a small boat moored alongside of us. They had found us with no radio or radar navigation. The driver of the small boat grabbed the gunwale and, as he did, accidentally dropped the handheld floodlight into the river. We started to sing again, adrift in the darkness.

We were found by two other boats, much later, and towed safely to the research station up the Raduga River, arriving after midnight.

In one of the dark log cabins of the station a meal had been prepared, borscht and a sliced meat with a spicy tomato relish. An open fire burned in a brick hearth.

The meal was cooked and served by a brown-haired woman who looked a bit like Ida. She smiled, showing two gold and three silver teeth.

Sasha came into the dining room and she showed us to our rooms.

It was Glubakovsky, white haired and distinguished, who led us in toasts, announced the day's activities, and circulated among us with a warm smile and kind words. He had brought several stern-looking friends who wore Adidas exercise suits and looked like part of the Russian mob. One of these fellows said to Markuu, "Good to meet you. I now have *three* friends from Norway, your king, your queen, and you." Did it matter that Markuu was from Finland?

At the first breakfast we were handed a booklet of printed abstracts of scientific papers, which the members of the International Society of Arctic Char Fanatics were scheduled to deliver during the course of our eight-day stay. In lieu of a paper, I would present a slide show of my 41st parallel travels. I knew the scientists would be interested in seeing photos of fish that they likely had only read of, the softmouth trout of Bosnia, *Salmothymus obtusirostris,* and the Mongolian grayling, *Thymallus breverostris.*

For the first several days, the talks, given two at breakfast and two after dinner, maintained a certain gravity. With vodka and late-night dancing, the seriousness deteriorated. Every morning fewer people showed up for the breakfast presentations. Then the speakers themselves failed to show. Markuu delivered his paper drunk. The scientists were taken by the clear cool air, the strength of the tall volcanoes, and the hearty meals of freshly caught fish. In the night they grabbed their towels, drank a few shots of vodka or whiskey, and walked in the dark under the stars to the *banya,* or Russian sauna.

Yoichi, Markuu, and I fed the fire with dry willow sticks and birch logs, which heated the stones on the opposite side of the wall. We undressed and stepped into the room of sour wood smells, dry and hot, and threw ladlefuls of water from a basin onto the hot stones to make steam. Sitting on the wood benches I could see the sweat beading up on my dry skin and rolling in droplets off my shoulder and chest. I rubbed the sweat on my skin and kneaded my face with my hands, taking deep breaths of the warm humid air.

Markuu spoke about the tradition of *banyas* in his country. "Many people have saunas in their homes in Finland," he said. "I take a sauna three times a week, if not more. The planks should have no knots in them because knots in wood get hotter and are unpleasant to sit on."

When we were at the point that we could no longer endure the heat, we stepped out of the sauna, ran down a path to the river, and jumped off the iron dock into the cold water.

During the day we made excursions to nearby lakes and tributaries of the Kamchatka River. I watched the Norwegians rig up their lines to fish in Lake Azerbache. A few tied on silvery spoons for char, others treble hooks, to snag the spawning sockeye salmon.

Yoichi and I walked up the small river over tracks of large bear. We caught many Dolly Varden char. The males, now in their spawning colors, were red like the south side of a ripe apple. I got to know Yoichi as we fished and hiked farther from the rest of the group.

He had left Kyoto, Japan, as a young man, traveled widely, and settled in Grenoble, France, where he worked at the university. He was short and thin and wore a leather pilot's hat given to him by the locally renowned Russian bush guide, Misha Skopets of Magadan. Igor Cheresnev told me that he had given Yoichi the boots and canvas jacket he wore, seven years before when Yoichi made his first trip to Russia. "He's spent more on patching them," Igor said, "than they're worth." Yoichi's first love was char but most of his scientific work was published on crayfish. Of all the people who had seen my slide presentation, he was the most intrigued by Johannes Schöffmann, who played a large part in the story I delivered.

"I would like to meet your Austrian friend Johannes," Yoichi said. "He sounds like a very interesting man. I am studying crayfish in the rivers of southern Austria, maybe I can visit him."

"You *would* like Johannes," I confirmed. "He is a fanatic in his own way."

"But more for *Salmo trutta,* the brown trout," Yoichi said. "Well, they are similar to char. I like how he catches them, by diving. It's very alternative. In Japan they have a way of fishing with a hammer. It's called hammer fishing, and it's best done at night. The fishermen hammer really hard on a stone in or on the edge of the stream and it stuns the trout lying underneath it. Then they net them. It only works, I think, with certain rocks, basalt maybe."

Yoichi kept three fish for a shore lunch, which we brought back to where the others were. The accumulated catch was piled on the sandy beach beside the lake. The women had made driftwood fires on which to cook the fish, burning it down to coals. I took a brief swim in the cold lake and then lay on top of some bear tracks, half sleeping, enjoying the sun and the smell of cooking fish. When I opened my eyes I saw Olga standing over me, her green eyes glaring. She was with her small square terrier that she called Cleopatra. In her hand was a hot cup of tea. I sat up to receive it. A light breeze lifted her wild curly reddish hair.

"James," was all she said, and kneeled down next to me. She looked at me with the adoring look a mother gives her son. Then she stood up and walked back to the fires to help cook the lunch.

That evening at dinner at the research station, in the dim of the wooden cabin, Dr. Glubakovsky suggested we have a banquet. He wished to recognize some individuals, present a special meal, and then get drunk and dance. Our friendly server who looked like Ida brought us a fish soup. Several cheeses were brought out, a berry wine, and salted salmon eggs. After dinner, those remaining moved the tables to clear a space for dancing. Russian dance music, rock and techno, played from a cassette player.

As the night wore on, Igor turned off the music and fetched a guitar.

The guitar was in poor condition and irreparably out of tune. He tried to play it but there was too much discord. Then a small middle-aged Russian man named Valérie, whose face reminded me of one of those black-and-white mug shots of an old-time train robber, came in carrying an amber-colored instrument, in tune and pleasant sounding.

Valérie played melancholy Russian war songs and everyone stayed to listen. My friend Fred told them I played some guitar and Valérie handed the instrument to me. I don't remember the first song I sang, maybe it was a Woody Guthrie tune. We all were very drunk.

Valérie, who now looked to me like a fox, was an exceptional folksinger. His songs were about war and love and mother Russia. Sergei Alexeev, another of the Russian biologists, crept up to me several times to tell me what the songs were about. "This song is for World War Two," he would say. "This is the song of the Russian Civil War." The drivers, the cooks, the boat mechanics, all the people who worked at Raduga Station, drank and enjoyed the music.

Olga danced to my songs, and when Valérie played I danced with Olga and held her by the waist and shoulders and felt the valley along her spine and put my leg between her legs and she laughed and let me go, spinning. I walked outside at one point into the chilly night to pee. When I returned I could no longer find Olga.

A small limping man, one of the boat mechanics who lived at camp and always seemed drunk, poured vodka in my mouth and corked it with a pickle. After the playing, around four in the morning, Yoichi and I went to the *banya* but the fire was out and the stones were not hot. We ran down to the iron dock by our cabin and jumped off into the river anyway. It felt good to be naked. When we came up dripping from the river, Olga was walking by with her dog, Cleopatra.

The next afternoon, a group of fifteen of us fished in a small tributary of the Kamchatka River, a short boat ride from camp. It was a kind of exploratory fishing trip and I caught a species of char I had previously seen only in Japan. It was *Salvelinus leucomanis,* the white-spotted char, locally known as the *kundja.* The first one I caught was small, cerulean blue on the sides with large round white spots. Yoichi caught a larger one, close to two feet long. We also caught Dolly Varden char, *Salvelinus malma,* and a char that Dr. Glubakovsky had first described, *Salvelinus albus.* No one else but Glubakovsky considered *albus* a separate species.

"It's identical to the Dolly Varden," Yoichi said. "It should be called Glubakovsky's ego char." We cooked fish for lunch on the bank of the river. The wind was down and the small biting flies were numerous.

Remnants of the evening redness could still be seen on the peaks of the volcanoes at the hour we returned to Raduga Station. In the dining cabin after we ate, the tables were cleared and moved for another night of festivities. Soon Valérie was plucking and picking his songs and singing with his throaty Dylanesque voice. I sat down on a bench in the corner and listened to Valérie until he had done playing and he handed me the guitar. The limping boat mechanic tilted a cup of vodka to my mouth and then fed me a pickle. Time passed, and individual members of the caravan slipped off to their cabins. I handed the guitar to Valérie and went outside to relieve myself. On the way back through the dark hallway into the room, someone took my hand. It was Olga. She led me into the kitchen, where we sat among pots and jars and fragrant bunches of dried sage and bay leaves. We lay down, embracing. Through a hole in the wall I could see the warm light of the dining room and hear Valérie's voice and the soft treble of the guitar strings on his fingers.

Olga kissed me and muttered several words in Russian.

"Sh, sh, sh," she said and kissed me again.

She led me out of the kitchen but someone was walking by the entrance so we ducked back in. Olga showed me that she would leave first and then I should follow. She made the moment darker than it was.

I met her in the damp grass under the full moon. We held hands again and she led me to her room.

"Sh, Cleopa," Olga whispered to her dog, closing her door. Cleopatra settled in a square bundle beside the bed. I watched Olga's dress fall as she stripped off her clothes. She helped me undress.

I touched her.

"Sh, sh, sh, James," she said. The cold world of wood walls, full moons, and wet grass was somewhere beyond the warmth I felt. I heard her delightful panting as I pulled her closer. I smelled salmon on her fingers. I asked Olga if I could light a candle and see her. Somehow she understood, for she moved the window curtain and moonlight spilled over her breasts. I saw the animal in her eyes.

Cleopatra was quiet and sleeping on the floor. Olga held me for a long time, then loosened her embrace and indicated to me that I mustn't fall asleep.

I kissed Olga's soft cheek and dressed and ran down the path through the dark, half elated, half bewildered, to the *banya.* I took off my clothes, hung them on a peg, and felt pleased to be alone in the dark. I moved a small washbasin below a large spigot and filled it with warm water, which I poured over myself.

The stones were still warm in the sauna. I put some water on them and light steam filled the room. I closed my eyes but did not sleep.

Later that morning, on my way to breakfast, I passed Cleopatra licking the dew off the grass. The dog had strayed from Olga's side but she must not be far. At midday we left Raduga Station and began the two-day trip back to Petropavlovsk.

Not until we arrived at the gate for our flights out of Kamchatka did I hold Olga again. She put a small figurine carved from mammoth ivory in my hand and said good-bye, smiling.

The other Americans and I were about to board the weekly flight to Anchorage, Alaska, when a tall man approached me and introduced himself. I recognized him from photos I had seen. It was Dr. Robert Behnke, my wellspring of information, the one who had introduced me to Johannes Schöffmann and shaped my travels along the 41st parallel.

"It is a pleasure to finally meet you," he said. "I'm Bob Behnke."

*The golden trout.*

JAMES PROSEK '02

"A pleasure for me too," I said. "It's so strange you're on this flight. I didn't know you'd be in Russia."

"I was supposed to join *your* party, in fact," he said, "but I decided to go on a separate research trip up the Amur River—some colleagues and I were studying Siberian taimen and lenok." He combed his lacy reddish hair with his large fingers. "I was determined to return with specimens," Behnke said, "but the vodka ran out and the bush guides drank all my ethyl alcohol. It was not an entirely successful trip, scientifically speaking. We could not navigate as far as we had hoped up the river because of a large log jam."

"You were fishing with nets?"

"Some, but I did better with my fly rod."

"Oh," I said. The wind was blowing hard.

"Why don't you visit me at the university in Fort Collins on your way home," Behnke said, pausing to stuff and light his pipe. The smoke blew across his face as he struggled to keep the match lit in the wind.

"I retired from teaching this year, my seventieth birthday, but I still have my office in Wagar Hall, the old veterinary building. I'll take you fishing on the upper Poudre River, it should be very near your parallel, forty-one degrees north. We have a lot of beaver ponds with brook trout; they are an introduced species so we can catch and keep some."

"That sounds great."

"We'll have a meal and share stories." He turned his back to the wind and puffed on his pipe. "Plan on coming about October fourth and staying a few days. I have a doctor's appointment on the third. You can coax me out of the office for some fishing. Where are you heading now?"

"I'm going to visit and fish with a friend in Berkeley."

"I did my Ph.D. work at UC Berkeley," he said. "Nice town."

When we arrived in Anchorage, making the four-hour flight across the Bering Sea, I said good-bye to Behnke and some of my new Fanatics friends. Then I boarded a plane for San Francisco.

## The Abridged *Schwarzfischer* Lexicon

Invitation to join ISOS,
the International Society of *Schwarzfischer*

President: Johannes Schöffmann
Vice President: James Prosek

Dear prospective members—Yoichi Machino, Dr. Robert Behnke, Pierre Affre

You have been chosen to become members of the ISOS because you exhibit characteristics of the founding members; that is, you foster the urge to fish by any means possible: 1. because you are a predator 2. in order to advance the world's awareness through art and science concerning the biodiversity of salmonid fishes, especially trout and char.

The following is a dictionary of terms and sayings of invention and collection in languages from countries where *Schwarzfischer(s)* have traveled in search of fish. It is a kind of *Schwarzfischer* code—*the Schwarzfischer lexicon.*

*Schwarzfischer*—German: literally, *black fisherman;* illegal angler, poacher, *pescador furtivo, pêcheur noir* (not proper to say *Schwarzfischer*s, but it is said nonetheless).

Spanish—official language of the *Schwarzfischer,* as it is the first language in which the *Schwarzfischers* communicated.

*"Mal vino es mejor que no vino"*—saying in Spanish of Johannes's invention: "Bad wine is better than no wine."

Gambatar—the name of an amicable driver in Mongolia used in

place of the Spanish infinitive *ir,* to go, and conjugated like a Spanish verb.

| | |
|---|---|
| *gambato* | *gambatamos* |
| *gambatas* | *gambatais* |
| *gambata* | *gambatan* |

*Schweinehund*—German: literally, *pig dog;* used as a light insult.

trout/char—favored fish of the *Schwarzfischer,* known variously as: *truite* (France); *trota* (Italy); *trucha* (Spain); *Forelle* (Germany); *forel* (Russia); *pestrofa* (Greece); *pstrimika* (Albania); *gouzelleh* (Iran); *ishkhan, bahtak, kharmrahait* (Armenia); *alabalik* (Turkey, Azerbaijan); *iwana, yamame, oshorokomo* (Japan); *bleikja, sjógen-gin, urrithi* (Iceland); *massi alé* (Kurdish).

Johannes-cut—antonym of shortcut, or a means of taking time off your travel by choosing a more direct route. Used ironically because a shortcut is intended to save time by abbreviating the route, but a Johannes-cut ends up making the trip longer.

cheese—v. smile. ex.: She is cheesing. Derived from American custom of saying "cheese" and smiling when a photo is taken. A Johannes slip.

*bekhwe*—Mongolian, roughly means there are none, a common response to questions asked in Central Asia; ex.: "Do you have any toilet paper?" "Toilet paper *bekhwe.*"

*yow*—Icelandic: yes. Repeated slowly in succession—*yow, yow, yow*—it suggests the American yeah, yeah, yeah. ex.: "Sure you caught a twenty-pound char—*yow, yow, yow.*"

*chishik unem*—Armenian: I have to pee.

*gola hanne*—Nepali method of fishing with explosives.

*Einfahrt-Ausfahrt-gutefahrt*—An amusing interlanguage wordplay. *Einfahrt* in German, *entrance; Ausfahrt—exit;* fart, American slang for flatulence; *ein* in German, *one; gute Fahrt* in German, good travel; good fart in English, good flatulence.

*fruta de tavuk—fruta de,* Spanish for fruit of; *tavuk,* Turkish for chicken. A convoluted way of saying "egg."

Oscar—a restaurant in Ulan Bator, Mongolia, with good, well-priced food. Used to describe any such rare place with those characteristics. Also conjugated like a Spanish verb, suggesting the following—"Let's go to the place with good, well-priced food."

| | |
|---|---|
| *osco* | *oscamos* |
| *oscas* | *oscais* |
| *osca* | *oscan* |

*pectopah*—reads *restoran* in the Cyrillic alphabet; i.e., *p* is pronounced *r.* But *Schwarzfischers* say this word with roman pronunciation, pronouncing it *pectopah,* to also mean restaurant, but in code. ex.: *"Oscamos a la pectopah,"* or, "Let's go to the restaurant with good, well-priced food."

*Sonnhof*—German: sun yard. The name of the bar in Sankt Veit, Austria (*Schwarzfischer* capital), where *Schwarzfischers* stop to have a drink on their way home from an expedition. The word connotes the physical bar, Sonnhof, but also an expression of faith, and an emblem of home. Usually expressed with a sigh of relief, *"Ah, Sonnhof,"* and joined with a *kleines* Gösser or Villacher beer.

*çanli*—Turkish for alive or live. Pronounced *chanli* (*ç* in Turkish is "ch" sound). Used in Turkish phrases such as *çanli muzik* (live music). Adopted by *Schwarzfischers* to describe a trout's condition, *çanli alabalik* (live trout)—opposite of *muerto.*

beer—*la bebida oficial del pescadores furtivos*—the official drink of the *Schwarzfischer,* known variously as: *cerveja* (Portugal); *cerveza* (Spain); *birra* (Italy); *Bier* (Germany); *birre* (Albania); *mpirra* (Greece); *bira* (Turkey); *pivo* (Russia); *garejur* (Armenia); *pichiu* (China); *shad airag* (Mongolia); *uTshwala* (Zulu).

Any additions to the *Schwarzfischer* lexicon are welcome.

# Golden Trout

I called on a friend to harbor me during my first days back in the States. I had last seen Greg on a trip we'd taken to climb Mount Shasta in northern California and wanted to go back to that area now to fish. Near the base of Shasta were several wonderful trout rivers on the 41st parallel.

I had been to Greg's house on Oxford Street in Berkeley before, but now I noticed how its decor had been carefully chosen to portray its inhabitant as an explorer. At the door I was greeted by K2 and Eiger, Greg's two Maine coon cats (named after famous mountains). The style of the furniture was African colonial, rattan, hemp, and leather, but a vegetarian version without the big game heads. On the walls were photographs from Shackleton's Antarctic expedition and his ship, the *Endurance.* Greg's shelves were stuffed with adventure books and expedition journals, the walls hung with ice axes. On shelves were photos of him and his friends, on Mount McKinley, on Everest, at the base of Machapuchari. He drove a Land Rover Defender to his daily job as a lawyer in San Francisco.

Greg and I spent the next several days on the 41st parallel in California, reliving memories of trips we'd taken together. North of San Francisco, through Napa and Sonoma, the country was dry and hot, amber and ochre, dotted with deep green oaks. For three days we fly-fished the upper Sacramento River near Mount Shasta for hard-fighting native rainbow trout, flattening pennies on the train rails, camping under the redwoods, sitting by roaring campfires under thick white stars. Then we traveled south to Yosemite National Park and hiked to the peak of Half Dome Mountain, overlooking the Merced River, which wound like a serpent through the valley floor. Greg gave me rudimentary climbing lessons on nearby rock faces and I contributed to his knowledge of fly-fishing.

We continued our trip southward to Death Valley and took the portal road near the base of Mount Whitney to a trailhead where paths began to wind toward meadow streams holding native golden trout. I had threatened to take this trip to Big Whitney Meadow so many times that I felt I had already been. On the hike up to Cottonwood Pass it all felt familiar, like I was revisiting one of my haunts at home. Perhaps I was nostalgic because the distant peak called Whitney was also the name of my first girlfriend.

It was twilight when we walked in among the redwoods and camped at the tree line on Cottonwood Pass, 11,600 feet. Before the sun set we could see Big Whitney Meadow below us and the faint glimmer of a ribbon of water. It got cold as soon as the sun went down. We made dinner in the vestibule of the tent, and were snug in our sleeping bags by the time night set in. I went to drink some water from my bottle but it had already frozen.

"It must be ten degrees Fahrenheit," Greg said. "But up here in late September you could get snow." Greg rolled over to sleep. "Thanks for giving me an excuse to take some time off from work," he said.

As we drifted off to sleep a peculiar thing happened. I became very paranoid and began to hear noises. I felt the ground shake and swore I saw a man through the tent standing over us holding a flashlight. He must be very cold, I thought. Why isn't he moving or speaking to us? An hour passed and still the illusion held. I tried to reach for my knife in my pack without making noise, and when I had it, I held it out, the blade open, waiting for the intruder to make a move. He never did.

"Isn't it always strange how the first rays of sun work to remove fears," I said to Greg the next morning. "I felt the ground shake last night, I know I did, it wasn't a dream, and I thought I saw a man through the tent holding a flashlight."

"Maybe you felt a tremor," Greg said, boiling water on a portable gas stove.

"Hey, I bet that's possible. There're a lot of tremors in California. But what about the man with the flashlight."

"Maybe it was the moon."

The grass underfoot was crunchy with ice, and the moisture in our boots was still frozen when we set off for Big Whitney Meadow with our fly rods. We packed our frozen water bottles and hoped they would melt.

The sun had spread across the low ground where we were headed. We moved swiftly down into a series of treeless soggy meadows where the sources of springs combined to make small brooks. In every tributary, every trickle, every finger that snaked up into the yellow grasses there were golden trout.

The trail came to water first beside a small pool maybe four feet across. I stood before it with my eyes fixed on the bottom and could easily see a half dozen fish lying above the gravel.

"Stand back," I cautioned Greg. "We don't want to spook the fish."

I strung my fly rod and rigged it with a small dry fly, trying to calm my excitement so I could make a good cast over the pool. When I did, the fly landed too perfectly; I saw no ripple, no wake, and no fish rising to take it.

"There's a film of ice," Greg noticed. "You see it? The stream froze over." The golden trout swam quietly and safely beneath.

We continued down the trail for a half hour waiting for the sun to grow stronger. The trail came to another brook, which my map called Stoke's Spring. All the springs in the meadow fed into Golden Trout Creek, which eventually ran into the little Kern River.

I caught my first golden dapping a dry fly in a two-foot-wide pool. From Greg's point of view several yards away, the pool of water was hidden in the tall grass and it must have looked like I was fishing for meadow voles. I lifted the trout out of the water and Greg came running to see it. I was astonished by its brilliance, the reds and yellows that seemed to have been burned into its sides by the California sun.

I proposed to Greg that we walk for one hour in search of bigger water, farther south where the multitude of stringers would join to make a larger stream. As we walked, Golden Trout Creek began to take shape and grew into a beautiful spring creek with a good even flow. The water was clear and I could see it was full of trout taking natural flies on the surface with abandon. The small trout took our flies recklessly, each a little jewel.

One fish among the many we caught outshone the rest in its brilliance. It was dark olive on the back with big indigo parr marks, a broad crimson band and yellow sides, a cadmium-red belly, a cerulean-blue rim on the jaw and over the eye, a big black iris the shape of a watermelon seed in a pupil of amber, the fins fine golden ochre and orange, and every color in its place and in the stream perfectly camouflaged.

When we stopped for lunch I took off my day pack and lay on a boulder by the stream with my face toward an all-blue sky, not a single cloud in it or anything, save a golden eagle passing from horizon to horizon. I had chosen to sit by a large pool for lunch; a cold breeze streamed through. In a small eddy of the creek between two big boulders, a half dozen golden trout between four and seven inches rose continuously to mayflies.

The whole earth was a meadow of golden grass with a thread of blue winding through.

"So you're heading east tomorrow," Greg said. "I just bought a small place in Colorado. It's a cabin on a hundred acres surrounded by BLM land. I've only stayed there once. I want to get out of here sometime and build a place there. You can stay, it's in the mountains and there's a small creek running through. There's supposed to be brook trout in it."

## Trout in the Nevada Desert

I rented a car and made my way from Berkeley, via Lake Tahoe's southern shore, into Nevada. I made it all the way to Elko, a casino strip in a desolation as vast as Central Asia's. In my room that night and into the morning, I could hear and feel the boom of trucks passing on Interstate 80. I slipped out of bed early and, following some information I had, headed up into the quiet desert in search of native trout.

I took to a gravel ranch road in Deeth up the dry bed of the Mary's River in the Humboldt basin. In a fisheries report published by the Nevada Fish and Game in the 1970s, the biologists detailed all the streams with native cutthroats left in Nevada. I had chosen one to fish called Wildcat Creek.

Skirting the rabbit brush at a good clip, a bristling yellow mass of flowers, a medium-sized bird hit my windshield. I was so startled that I spun off the road into a dry ditch. The front bumper was a little scuffed and I dropped hot coffee on my lap, but I managed to muscle the car out and things seemed to work fine.

I thought it may have been a kestrel or a whippoorwill and spent some minutes scavenging the roadside looking for it. It was a whippoorwill and I pulled a couple wing and breast feathers, and thought someday I might tie a fly with them.

On my map I had marked Wildcat Creek; it lay on the fringe of the Humboldt National Forest. I thought I was getting close but saw no forest. Ahead of me was parched ground and sagebrush. I saw a man standing beside a pickup truck and stopped to ask directions.

"Mornin'," he said.

"Hi. I was looking for a creek called Wildcat," I said. "I was thinking of doing some fly-fishing there. You know where it is?"

"I should, I own it and the several thousand acres around it. My name's Bill Gibbs." He shook my hand. "There should be trout still, but I was tending young cows soon to be shipped off to feed lots in Nebraska the other day and saw that the fire'd burned over it. I don't know if the trout survived.

"You'd better not go up this way," he said, looking at my dusty compact car. "It'd be difficult to get to Wildcat in that—the road's washed out. If you go back north up the Deeth County road you'll come to a small dirt road on the left just past the next cattle grate. Take that as far as you can and then you'll have to walk at least two miles before you get to any water. It's been a real dry summer and like I said, the fire jumped the creek but a few might've made it."

"So," I said, leaving *Schwarzfischer* protocol behind. "I can fish it?"

"Sure," he said.

I thanked Bill Gibbs, made for the cattle grate, and took the first left, flushing rabbits, crushing ants, through sage and rabbit brush, until I got to where the fire had burned and the road became impassable.

I made my way on foot through the burned country. The brush was standing charcoal and drew patterns in black lines on my calves as I walked. The sun was intense. I spooked a dozen antelope from a water hole. There was a basin between the hills. I thought the creek must be there; then in the distance I saw a lone green aspen and some willows. Somehow they had been left untouched by the fire.

As long as I had walked the willows didn't seem to be getting any closer. Was this some kind of trick? And then I was there, staring into a thin ribbon of water that barely flowed.

Looking carefully into a pool, I saw a small trout that I recognized as a cutthroat. It was a different shade of olive from the algae and detritus on the bottom and was sparsely spotted with round black points.

I strung up my fly rod, but only the front half, and dropped a nymph in the pool. As soon as I did, the trout took it and I was holding it in my hands, admiring its Tuscan red sides, its violet cheek, and its look of wildness. I thought of a discussion I had had with Johannes about how you knew a native trout when you saw it. Everything just seemed to make sense; the trout was not only a fish, but all the colors of the land around it. Its spots were charcoal black and its sides the color of a desert sunset. It had survived the fire.

Content, pleased, lost, I drove eastward on the highway through the largely desert land. Soon I was in the Bonneville Salt Flats and then on the south shore of the Great Salt Lake in Utah. It was so like the inland seas of Asia, backlit by dust devils, vast and desolate. I had once traveled, years before, around Utah with a friend from high school who was Mormon. We had borrowed his grandfather's truck in Toole (pronounced *Towilla*), a small town near Salt Lake, and spent the trip fishing for trout and sleeping in the homes of his relatives. Saying his name in my mind, Josh Blackwelder, conjured the warm backyards of his relatives' homes, plum trees hung with ripe fruit, goats tethered to stakes in the grass, the earth dry and fragrant, blond Mormon girls with eyes like golden raisins, prospects of a different reality, of religious hallucinations from the intense heat.

Last I heard Josh had joined the navy. I spent the night in Provo.

# THE CABIN

I'll eat anything that doesn't eat me first," the man at the market said to me.

I had asked him what the dead rattlesnakes hanging outside his shop were for.

I bought canned beans there, rice, and packaged sliced bread. I enjoyed buying provisions, whether it was in a supermarket before a snowstorm or a country store before a camping trip.

The backcountry cabin Greg owned was nestled in a private spot in the center of BLM land in northeast Colorado. It was an old homesteader's cabin on a hundred acres, which he hoped to one day expand on and winterize. For the time being it was a nice place to crash on hiking and fishing trips or to offer to his friends.

I followed Greg's directions off Route 40 to a dirt road, by the hunched bodies of charcoal-colored bison. Some way down the road I saw several elk and began to think I was on a safari. At the end of Greg's directions, on top of a small rise, dark against the golden grasses of autumn, was a cabin.

The cabin was made of lodgepole pine and the interior, even after I had found the key, gone in, and opened the shutters, was dark. There were two beds in one room beside a wood-burning stove and in the other, smaller room there was a wood-burning cooking stove and a shelf with two cooking pans.

"There's split wood out back to burn in the stove," I remembered Greg telling me. Nailed to the inside of the kitchen door, I'm sure from a previous owner, was a note:

> To all visitors: there are fresh sheets hanging in a sack here so the mice can't get to them. You can make your own bed and when you leave put the sheets in the other sack with the used sheets. There's some food too, as you can see, but only use it when necessary. There are plenty of trout in the creek.

Two out of three promises had not been kept. There was no food and no fresh sheets. I hoped there were trout in the creek.

Toward dusk, when I had settled in, I sat on the front porch of the cabin in my underwear and heard a coyote yapping; then several more yelped with the first. I scanned the hills, now a deep rose color, and in the last light of day I saw them racing in formation.

It took some effort to find matches to light the gas lamp.

"Shshsh hississississ shshsh shshsh Jamesamesj mesja esjam sjame sh sh sh," it said. It seemed to speak to me, and I took some whiskey that I'd brought and drank it, and that helped me hear. Maybe I wasn't alone. And because it was dark outside and the only light for miles around was emanating from the interior of my cabin, I had ignited a beacon for all predators. I feared to attract attention, I feared the voice, and I shut the flame from its fuel source.

Darkness pervaded the room.

I found my bed and the cold sleeping bag. I should have worn my flannels, I thought. I'll keep the flashlight by my bed and my knife too. My head was on the pillow, there was a pillow, the bedsprings creaking as I settled in; the smell of sage carried in a cold air through the open window where I had placed a screen to keep out the bugs. I could hear the creek below me, could imagine reflections swirling in its currents, and then I heard a noise. What was that noise? What the hell was that noise! It was a mouse scuttling—no. It was not "ta ta ta ta ta ta," it was "*thump,* drag, *thump,* drag, *thump,* drag," and it sounded as though it were on the floor of my cabin, approaching my bed.

If you were to judge an animal's gait and size by the interval between its steps—long—this animal was large. Had it come through the only open window? If so I didn't hear the screen fall to the floor. I'll reach down for my flashlight and lie still, I thought, holding my knife at the ready. I expected to see large yellow eyes reflecting in the light when I turned it on, but there was nothing, the sound had stopped. I got out of bed and went into the kitchen to pull the screen out of the open window and close it shut, but nothing was there. I put my head out and looked up at the stars. They were prominent and bright. It was not until I got back into bed and listened for the steps to start again that I realized the sound was coming from above me, in the attic. I dared not open the hatch to the attic and look, so I went to sleep, thinking the animal would not open it either.

When morning came, after sleeping the dark hours to dawn, I lit the woodstove with the summer stock-market tables. The heat produced from capitalism burning expanded the stovepipe and made a percussive boom. On the face of the stove was written COLE MFG. CO. CHICAGO ILL.

I heated some water to make oatmeal and hot chocolate and pretended I was a frontiersman. The heat from the stove became oppressive and I had to back off from it. Maybe I was in a *ger* camp again on the fringe of the Gobi Desert. The driver Gambatar had already come into the low dark tent to light the stove. Or I was a soldier then in the army of Genghis Khan somewhere on the desolate steppes, fighting for an empire in a desert. Empires were just illusions. Battles had been fought in places where rivers ran and cows grazed; at the battles' ends, such places were peaceful again—Gettysburg, Normandy, a beet field on the Elbe. What about "the dreadful but quiet war of organic beings, going on in the peaceful woods, and smiling fields" that Darwin wrote about in his journal in 1839.

I stepped out on the porch in my long underwear to survey the yellow hills. My breath drew out in a smoky plume and I embraced myself in the cold. It was that precise time in the morning when the stars begin to disappear and a blueness pervades from horizon to horizon. Before me the sun was rising but its warmth had not yet reached the ground. I turned around and looked at the cabin, its front door glowing in the soft light, and then looked above the door to the small triangle of attic space beneath the wood-shingled roof.

Thinking the attic was too small to accommodate the animal I had heard the night before, I walked around the cabin to see if there was a hole somewhere big enough to justify my creature's entry. Finding there was not, I reasoned that the animal had found its way in as an infant and grown up there, which of course was absurd. During the four nights I spent in the cabin, I never heard it stir as long as the sun was out, but after dark, without fail, it began to move. Was it a bat? If so it was porcupine-sized at least.

Soon the western sun, the same sun I'd seen everywhere else, was high and warm and my early fire in the stove burned to embers. I was standing on the porch preparing my fly rod, thinking about where I should start my fishing. The scene before me was close to my mental images, as I had never been, of the African savannah. From anywhere amidst the tall grasses, dry and golden, I was prepared to see the sandy shoulder of a lioness.

Down in the meadow sloughs of Upper Blacktail Creek, as the creek was called, there were water buffalo, though really they were bison, and on the hills above were scavenging hyenas, though really they were coyotes. The roads along which moose and elk can sometimes be seen were beyond sight, no trails led to this spot. I was alone, in my khaki fatigues, bracing my over and under, smiling and squinting in the dry sun, taking my emerging mustache with thumb and forefinger, contemplating the creek winding through the willows and sage.

As soon as I walked off the porch into the grasses, I noticed that every plant had its unique way of hitching seeds to my pant legs, socks, and boot laces. I carried them some distance up Blacktail Creek and then circled toward where the sun had risen and walked up a hill from where I could see much of the area, and my cabin. At the crest of the hill there was a massive boulder that must have been ten feet high. I wanted to be on top of it, so I sank my fingers into the rock and pulled myself up. I lay facing the sky, watching the sun emerge from and disappear behind large white clouds. After a while, the wind picked up and two small raindrops fell on my cheek. The air became cool and I was very hungry, as I had eaten only a bit of oatmeal that morning. I got down from the boulder and returned to the creek to fish.

I have always found it difficult to fish on an empty stomach, and the wind was making it difficult for me to cast the fly in the water and not in the grass. After fishing four bends in the creek with a dry fly, I had not seen or caught anything. Then I came to a large pool with a small cascade at the head of it. In that pool were many brook trout, and before long I had two strung through the gills on a willow stick. It was early afternoon and I returned to the cabin to cook them.

They were females, and each had a pair of golden masses of roe. Still speckled, like the nighttime sky, I laid them on the kitchen table, filled a skillet with oil, and lit the woodstove. After about half an hour I found I could not get the stove hot enough to cook on, so I lit a small open fire behind the cabin and roasted the trout on willow skewers like marshmallows. I ate them like a cob of corn; their flesh was sweet, as if they had been marinated in sugar water. I licked the willow stick clean and put out the fire. Then I took out a book and read on the porch with a glass of whiskey. I sampled the book and sipped the whiskey until the light grew dim and the first star (probably a planet) showed in the sky. By and by, the creature walked again in the attic.

I still could offer no explanation for it and crept slowly to my bed. Shortly, I fell to sleep.

When I woke the sun had already risen. I lit the woodstove and sat by it. The sun beamed strongly, and I moved a chair out onto the porch to sit and listen to the creek. When the sun was high enough

*Le grand bécard vainqueur.*

HURTU
J. PROSEK '02

to reach the water, I undressed and walked barefoot through the sage to bathe in a deep pool of the creek where I had caught the fish. There the water was deep enough to cover my body. I could not take the cold water for very long and I stood up, climbing onto the grassy bank, returning to the cabin to dry off. I put on jeans, and my khaki safari shirt, and sat in a chair on the porch.

I tromped again that day through the golden hills and valleys. At dark, the nameless creature walked in the attic.

## Visiting Dr. Behnke

On October third I drove along Route 40 toward Fort Collins, Colorado. Groves of aspen, nestled in the crooks of hills, were changing to yellow and orange like bowls of ripening mangoes. Eastward toward the Continental Divide there was not a breath of wind. Just before dark every lake was still. The pristine reflections of the changing aspen were disturbed only by the rings of rising trout.

At Rabbit Ears Pass I drove in minutes from the drainage of the Colorado River and the Pacific to that of the North Platte and the Atlantic. Night came completely as there was no moon, but I could hear the rushing currents of the Cache la Poudre River out of my open window. I saw no sign of anyone on the narrow road until I came to a roadside bar and heard music. It was filled with college students from the state university. Some miles beyond I came to the town of Fort Collins.

Behnke lived with his wife, Peggy, in a modest ranch house on East Prospect Street, just outside of town and the campus of Colorado State University. I arrived there the next morning and was greeted by Bob Behnke at the door. He smelled of sweet pipe smoke. "Oh, James, come in," he said in a slim, nasally voice. "Did you have any trouble finding my place?"

In the living room, the first room on the left, there was nothing that resembled fish, but the hallway was stuffed with renderings of trout in all media. There were more paintings of trout one floor below in the television room, and in the adjacent kitchen. It was a well-used kitchen with full spice racks, dried sage, rosemary, and flowers hanging from a beam, liqueurs with sticky bottles, some blackened pots, and a gas stove. I sensed that if Behnke were the true *Schwarzfischer* I expected him to be, he would hold cooking, eating, and drinking wine on a level with his fishing.

We exited the house by a door in the kitchen and Behnke showed me his yard. Beside the house was a red barn where he kept a mule, and beyond the mule was the garden where he and his wife grew tomatoes and peppers, squash, sage, rosemary, and several varieties of mint. Beside the garden were small fruit trees, and on some there were ripe apples. The yard sloped down behind the garden to a pair of small ponds thickly lined with willow and cottonwood. Three white geese honked at us and the mule by the barn brayed in response. Behnke stared into the ponds.

"Before these willows and Russian olive grew up and sucked all the water out, I had trout in this spring hole—some Snake River cutthroats up to twenty-three inches." Behnke talked continuously in his nasally voice, as if he were trying to share everything he knew. He touched his touseled reddish hair. "The Russian olive is a pretty tree," he said, looking at one, touching its leaves and taking out his pipe. "It's introduced but I like them, they provide a lot of cover for the birds." He paused to stuff tobacco in his pipe and light it. "I've got a special spot for us to fish this afternoon," he said.

Later that morning, Peggy joined Behnke and me on a pond near the town of Nederland, east of and above Boulder. The pond was a sort of fishing club that Behnke and some friends had organized and stocked with greenback cutthroat trout. The greenback cutthroat, native to headwaters of the North Platte and Poudre rivers, was thought to have been extinct by the late 1960s, but Behnke did not believe that to be so. As a scientist-angler he set out to survey some very remote streams with a fly rod hoping to find some. He did and was instrumental in establishing a recovery program, reintroducing greenbacks to much of their native range.

We fished in the pond at Nederland from an aluminum boat while Peggy sat on shore. Behnke stripped streamer flies quickly through the clear water, catching several greenback trout and pointing out their distinguishing characteristics. He held the fish as if it were the first he had seen and then let it go. He stopped to stuff his pipe and then suggested we return to the bank to eat.

Behnke had prepared a gourmet lunch of roasted poblano peppers stuffed with prosciutto and cheddar cheese, premixed margaritas carried in mason jars, another jar with homemade gazpacho soup, and corn tortilla chips. Peggy and I sat on wooden benches at a picnic table and Behnke laid the food before us. He poured me a margarita and the three of us ate and drank and talked. I had remembered reading in an interview in *Fly Rod & Reel* magazine that Behnke was born in Stamford, Connecticut. I was born in the same town and over lunch I asked him about Stamford.

He began talking of his childhood and his early fascination with fish. "My youth was spent much like yours," Behnke said, "chasing wild brook trout in small streams.

"Ever since I was a little kid—I was fascinated looking at fish," he said to me, cutting one of his stuffed peppers and taking a bite. "I started catching pumpkinseed sunfish in local ponds. Then I caught my first brook trout above a small dam on the Rippowam River when I was eleven. I ran home with it and put it in a tub of water. I was observing it and observing it, and I got so admiring of it I was going to take it and put it back. But I went in to eat lunch, and it jumped out and killed itself, so I ate it."

Behnke read all the books on fish in the Stamford Library and his "obsession with fish," as he called it, grew. He took up fly-fishing, and at eighteen he went to work for the Yale and Towne Hardware Company, like his father and half the people in his town, making locks. "One thing I looked forward to each summer if business was slow, I'd be laid off and go on unemployment for a few weeks and go fishing all the time," he said. In 1952 he was drafted into the army. In the spring of 1952, stationed in Japan, he spent his off-duty hours fishing for the native char, *iwana* and *yamame,* in small mountain streams. When he returned home from the army the Yale and Towne factory had closed and moved south and he had no job. He went to a counseling service in New Haven that helped veterans find jobs. They gave him some tests. " 'Hey, you should go to college,' they told me, but I didn't know what I wanted to study. The counselor asked me what interested me and I said, 'Fish.' I had never dreamed it was possible." Behnke studied ichthyology at the University of Connecticut and graduated with a bachelor's degree in three years. As an undergraduate he read fisheries journals nights and weekends and published the first paper on the freshwater fishes of Connecticut since 1844. The star fisheries student received offers from several graduate schools but it was a call from Paul Needham, a biology professor at the University of California, Berkeley, that led to his future as a preeminent trout scientist. Needham was doing a study on the trout of the western United States, planning to drive from California to Alaska collecting trout, with a fly rod, and he needed a research assistant. He offered the job to Behnke and they traveled together for several months in a pickup truck across western North America searching for trout.

Behnke did his master's thesis on the Lahontan cutthroat trout of Nevada, and then got a grant to study the Salmonidae of the world for his Ph.D. He spent months researching museums and streams in Britain, Russia, and Yugoslavia and received his Ph.D. in 1964. He taught for several years at Berkeley, filling Paul Needham's position when he died, and then settled in as a professor of fisheries and wildlife biology at the University of Colorado, Fort Collins.

The sky was a brilliant blue. It was a beautiful day by the lake for a picnic and for fishing. After lunch there was a hatch of small sedge and we caught one trout each on dry flies.

In the days I spent with Behnke, returning to fish the pond with greenbacks near Nederland, he was curious about what I'd discovered in my travels, what the countryside looked like, the trout and the rivers. But most of all, he was interested in Johannes Schöffmann, the mysterious amateur trout scientist he'd been in correspondence with for years, and whom he'd never met face-to-face.

"You say he's a baker?—fascinating. I've only known a few like him, an amateur who is as effective or more so than most professionals. Sometimes, you see, the amateur is more efficient because his conclusions are born from his observations and not the other way around.

"I first heard from Johann [as he sometimes referred to him] in 1986. He had written me with information on two very remote and rare species of brown trout, *Salmo platycephalus* of Turkey and *Salmo pallaryi* of Morocco. He determined that *pallaryi* was extinct and that *platycephalus* was vulnerable to overfishing, writing up his conclusions and publishing them in an Austrian fisheries journal. When I read Johannes's scientific papers, the range and depth of knowledge displayed indicated formal training in ichthyology. Now that I know he is a baker, his level of expertise is all the more impressive."

Though Behnke was no longer teaching, he still kept his office in the basement of the old veterinary medicine building, Wagar Hall, on the campus of CSU, and I expressed a wish to visit it. My secret wish was to have a week to pore over his files, read papers and letters and books and look at specimens, but I would not press such an agenda on this trip. On one of the sunny afternoons I spent in Fort Collins we walked across campus to see his office.

[illegible] was just as I had pictured the office of a world-class trout tax- [illegible]t, piled with books, photos, and papers, and no larger than [illegible]k-in closet. He had a separate examination room where jars of preserved specimens—fish he'd collected over forty years—stood on tall shelves. He took me into the specimen room and we walked among the fish. By a small sink and table where he performed his examinations there was a refrigerator. In it he kept specimens he was examining at that moment, but it was also stocked with jars of his homemade gazpacho soup. It seemed he took his gazpacho as seriously as his taxonomy. He took out a jar and poured a cupful for me, handing me a spoon.

"I taste it and add ingredients until it suits me," he said. "This batch is two weeks into the tasting process—it sometimes takes me a month to finish a batch. I start with a base of tomatoes and cucumber ground up in a blender, then I add any number of the following ingredients: cilantro, cumin, garlic, lime, onion, basil, habañero tabasco, sour cream, balsamic vinegar, and poblano peppers." He showed me around the lab as I ate the gazpacho. "You like it?" he asked.

Among the fish, motionless in the bottoms of jars, were specimens Johannes and I had collected in Turkey and Central Asia and mailed to him from Austria.

Back in his office, Behnke gave me papers on the trout of Lake Ohrid, Macedonia, and Lake Sevan, Armenia, of which he had duplicates. He showed me a beautiful old book with color plates of the fishes of Japan and then sat down in his reclining chair behind his desk to light his pipe. We made plans for fishing the next day, my last in Colorado before I returned home. Behnke suggested as we were both Connecticut Yankees that we should go up to a tributary of the Poudre near Cameron Pass and fish for brook trout (which were not native to Colorado but introduced from the east). The trout were plentiful and we could keep some to eat for dinner.

"I don't mind harvesting introduced trout," he said. "Even though they are wild and beautiful."

The next day we drove up Route 14 toward Cameron Pass to Behnke's fishing spot. Behnke included me that day in an annual ritual of his and Peggy's, to go up on the pass and drink a bottle of inexpensive champagne while watching the autumn colors on the aspen leaves.

"Well, I suppose this is what retirement is supposed to be like," Behnke said as he lowered himself carefully to sit in a meadow by a series of ponds that beavers had dammed in the creek. I sat next to him. He lit his pipe and poured more champagne in my cup. It was a gorgeous day, the sky was deep blue, the water in the creek was black, and the aspen in the crooks of the hills were golden yellow. "What do you say we catch a half dozen small trout for supper?" Behnke said.

I watched him string up his fly rod and tie on a small bead-head caddis nymph. He walked to the soggy bank, made several casts, and hooked a brook trout. Smiling and laughing Behnke pulled the trout onto the bank, pouncing on it in the grasses. He then unhooked it and put it in his creel.

"Take a look in the shallow end of some of the beaver ponds and you can see the trout spawning," Behnke said. The females were digging redds by sweeping silt from the gravel. The males were waiting to fertilize the eggs that would be laid there. Purple clouds spread across the sky and their reflection made it harder to spy on the trout.

I fished in the deep parts of the ponds and caught four trout, which I strung on a forked willow branch. They were small and delicate, though their delicacy was partly an illusion, as trout evolved in the most indelicate of geological circumstances. I thought how peculiar it was that trout buried their eggs in gravel and were born from beneath the stones in the river bottom. When I returned to Behnke's side an hour later, he was sitting in the grass with a half dozen brook trout, already speaking of how he meant to prepare them.

"I'm going to make a kind of tempura of squash blossoms and okra and serve it with the brookies—I'll stuff their cavities with mint and sautée them in pumpkin oil with a little garlic." He turned his pipe upside down and tapped it on his waders. Then he stuffed it with fresh tobacco from a small pouch. We sat down to admire the landscape.

"Some dark clouds moving in," Behnke observed.

"It's nice up here," I said, lying back in the grass and sage, feeling the cold clean air. Behnke contemplated the brook trout lying on the dry grass next to his fly rod.

"I think we need to pursue and kill—as *Homo sapiens,*" he said. "It's part of what we are. Indeed, we are predators." He lit his pipe. "You know, I never realized until you pointed it out to me that I'd spent most of my life in two distant cities on the same latitude," he said. "Stamford, Connecticut, and Fort Collins, Colorado, are both on forty-one degrees north. You chose a good parallel to fish; it may very nearly be the best in the world for diversity of native trout and char."

That night we ate the trout in Behnke's house and drank chardonnay, listening to a recording of Schubert's *Trout* (*Die Forelle*) play in the background. While we were drinking a favorite aperitif of Behnke's, a Chilean liquor called *pisco,* a storm was blowing over the pass where we had fished that day, dropping a foot of snow and turning the aspen leaves black.

# Home

Denver is not a distant drive from New York if you set your mind to it. Friends of mine have made the trip in twenty-five hours, stopping only for gas. I had no pressing reason to get home, but its gravity pulled me.

Driving across Nebraska, where the hills at last had become plains, a magpie flew from where it had been feeding on a road-killed deer. The stench of feed lots kept me awake as I drove through broad oceanic fields, cows' breaths visible and plaster white in the cold day. I crossed the Missouri River and stared again into black water. Then I crossed the Mississippi and rode through Illinois. I did not stray from the nondescript highway until I came to Indiana, ending up at night in Evansville. The small brick town was lit by streetlamps, which reminded me oddly of Paris.

I ate breakfast the next morning at a small diner beside the Ohio River. Barges, bright and huge, labored against the current. Blue-haired women played bridge at a table by the window. Perhaps it was the painter John James Audubon who led me here. He had spent some time nearby, living, failing at business, and painting birds in Henderson, Kentucky. I spoke only to waitresses, and my memories were jumbled in my head. There is no clarity—through Louisville, Cincinnati, Columbus, Zanesville, and on into Pennsylvania.

"That's quite a sculpture," my father said, looking at François's masterpiece, the *grand bécard vainqueur.* But I was dreaming this. I had fallen asleep on the road and caught myself just in time. In my mind the leaves on the sugar maples had turned color and some were falling as we spoke. My father's hair had become whiter, but I did not find the changes in his face so noticeable at first. "I want to hear about it all," he said, "but wait until morning."

It's one of the first times in my life I took his advice. I pulled off Interstate 70 and got a room in a motel.

I took off my clothes and got into bed. I shifted to look out the window at the moon. Home revealed to me what I had secretly feared, that despite having traveled bleak countries, when I woke in the morning I would be only who I was when I'd left.

I crossed the Delaware, in my mind again, as if the fog had lifted from my brain. I found greater New York and its energy intimidating and foreign. It was too dark to see the Hudson when I crossed the George Washington Bridge, but I could see lights from the city and the cables of the bridge, and an hour from there I was beyond the glow of New York. The lights went out and I was on the dark winding rural roads near my home.

Then sleep. There was a place, a house in a mountain village roofed with stone shingles shaped like fish scales. Below it was a small pond. It could have been Portugal, it could have been Japan, but it was not, it was my home. I dressed and walked through a cold air that smelled of woodsmoke. Through the leafless trees I saw the pond near my house where I grew up fishing. It had three inches of clear black ice on it. I walked out onto the ice, so clear it felt as though I was walking on the water itself. I lay on the ice, facedown, and shielded the light from above with my arms, peering into the world beneath the surface. The pond is still, I thought, but through the window of ice I saw a mysterious current beneath, one that moved the weeds, and carried small scurrying organisms. Then I felt a cold wetness on my bare neck, and by the time I lifted my head a small white film of snow had concealed the black ice and the window to that other world was closed.

# Afterword

## A Little-known Twentieth-century Artist Named Iliazd and His 41°

Few individuals are aware of the geographic parallel on which they live, and fewer have considered whether they are on that parallel for a reason. One man constructed a life philosophy around it—Ilia Zdanevitch, born in 1894, in Tbilisi, Georgia (the former Soviet republic), on the 41st parallel.

At seventeen Ilia Zdanevitch moved to St. Petersburg to study law but soon gave up his studies to paint and write poetry, assuming the name Iliazd, a combination of his first and last names. Through his brother Kiril, who also lived in St. Petersburg, Iliazd met the futurist painters Victor Barthe and Mikhail Ledanter. He became part of their avant-garde circles, continuing to live in that city as a confirmed futurist for six years, developing his modernist ideas.

Iliazd returned to his home city of Tbilisi in 1917 and settled in the Caucasus Mountains working as an apprentice to a publisher. Shortly thereafter, with the help of two poet friends, Iliazd founded a small modernist magazine intended to house the literary experiments of himself and his friends (including writings in a language of their own invention called *zaum,* which advocated a redefinition of language based on word sounds). They named their magazine the *41st Degree,* a reference to the latitude of Iliazd's home.

In Iliazd's mind, the 41st latitude seemed to connect his small Georgian town of Tbilisi with a culturally rich and politically powerful world abroad. "It is at forty-one degrees," Iliazd wrote, "that most of the great cities of light are located—Madrid, Rome, Constantinople, Beijing, and New York." Numerologically speaking, 41° harbored other relevance to Iliazd. "Jesus remained in the desert forty days and forty nights," he wrote, "and on the forty-first came out cleansed and emerged stronger. Forty-one degrees Celsius is the body temperature at which a feverish delerium takes over the body and we die."

In 1921, despite such strong feelings for nationalism concerning his native latitude, Iliazd moved away again, this time to Paris. He was known later for saying that "although Paris was not on the forty-first parallel (it's on the forty-ninth), it should have been." Now, at the age of twenty-seven, his plan was to impose his futurist ideas (language of *zaum* and avant-garde use of typography in printing) on Parisian and expatriate artists by establishing what he fantasized about calling the University of the 41st Degree. His vision for the university was as

> A society for the building and exploitation of the world's political ideas—Peking, Samarkand, Tbilisi, Constantinople, Rome, Madrid, and New York. Sections at: Paris, London, Berlin, Moscow, Tokyo, Los Angeles, Teheran, and Calcutta. Universities producing books, newspapers, and plays useful for the progress of the idiot literate. 41° is the most powerful organization in the van of the avant-garde in the field of poetic industry. Its beginnings go back to the first decade of this century when, thanks to the work of its collaborators and pioneers, there were discovered in various parts of the terrestrial globe extremely rich and unexplored areas of language. At the present time, 41° embraces more than sixty linguistic systems, including new territories, and attracts new capital with each succeeding year.

But Iliazd's attempts to start the University of the 41st Degree failed. What ensued for him was a creative desert that lasted eighteen years. In 1940 Iliazd emerged as a printer of art books of some renown, creating the works that he is best known for today.

Through 1974, Iliazd published twenty *livres de peintres,* painters' books, each in editions of fewer than one hundred. These books, illustrated by a handful of preeminent artists whom Iliazd had courted as collaborators—Picasso, Miró, Giacometti, Matisse, Max Ernst, and Jacques Villon—conveyed their subjects not only through the meaning of the text but through their typography, illustrations, and materials. In these books he had finally accomplished in print, at least in part, a few of the ideals he had created for himself to live by.

In the words of Françoise Le Gris-Bergmann, Iliazd's view of the book was "both as an object and a receptacle, as a site as well as a stage, as an emanation of the Word as well as of a kind of choreographic imaging."[1] Ultimately, however, Iliazd's vision of the perfect printed book may have exceeded the possibilities of the tactile world.

His standards for production were uncompromising, he wanted total control, was idealistic and noncommercial. It was a feat for him to find willing collaborators at all, but he did. "He searched for the perfect paper stock as though he was hunting for treasure," wrote Audrey Isselbacher, "and sometimes invented his own typography to best convey the meaning of the text. Always his subject is the unknown artist or poet, the nobody."[2]

In his book *Le Frère Mendiant,* a tale of a voyage through Africa by an anonymous fourteenth-century Franciscan monk, Iliazd chose materials that he thought would best express a narrative of geographic exploration and movement in order to create a harmony between the book's subject and its appearance. "On any page of *Le Frère Mendiant,*" says Bergmann, "we can read the sinuous outlines of the coast of Africa through the initial letters of each line,

[1] *Iliazd and the Illustrated Book.* New York: The Museum of Modern Art, 1987.
[2] Ibid.

their slight or radical unevenness creating for the navigating eye the entrances to grottoes, lagoons, steeply rising cliffs—and, beyond, the vast spaces of the horizon and the sea. Paragraphs, indentations, the unevenness and gaps of the lines—all constitute the cartography of the page, its geomorphology, the material of a spatial and topical representation."

Iliazd not only created unique books, he also made elaborate containers, covers, folders, envelopes, and slipcases in several layers of paper to create the atmosphere similar to a stage curtain, to veil what was within. The anticipation of the book for Iliazd was foreplay. "At times," wrote Isselbacher, "the dramatic quality of his volume's architecture is contextually relevant, as in the narrow verticle format of *La Maigre,* a biting satire on the vanity of a thin woman written by Arian de Monluc in 1630. Its stiff parchment cover is impossible to open—one must remove the leaves to read them—and even the fibrous folder around the parchment is rough and dry to the touch, like the brittle and rigid character so vividly described by Monluc."

Each of Iliazd's books, through a harmony of text, image, and material, was meant to be a world in itself, a landscape, a geographical site, and always that site was marked with Iliazd's imprint—41°. The forty-first parallel in Iliazd's work was not only a geographic location, but also a philosophical concept.

## THANKS TO:

johannes s., ida s., pierre a., jennifer p., julia h., françois c., marie-annick d., andré s., philippe b., larry a., elaine m., joe d., krista s., bob b., steve p., vincent g., judith s., hill a., whitney t., joe h., jim mo., jim mu., greg m., louis p., lynn p., kristina h., terry h., etay z., david t., bob c., steve s., taylor h., harold b., maria h., valerie g., ryan r., greg b., nick l., agnes p., kevin d., monte b., and all others who helped with my travels, contributed to my learning, shared their enjoyment of life, or aided with the manuscript. brazil wins today!